Economic Development in Central Asia

Emerging Markets Studies

Edited by Joachim Ahrens, Alexander Ebner, Herman W. Hoen,
Bernhard Seliger and Ralph Michael Wrobel

Vol. 6

Joachim Ahrens / Herman W. Hoen (eds.)

Economic Development in Central Asia
Institutional Underpinnings of Factor Markets

Bibliographic Information published by the Deutsche Nationalbibliothek
The Deutsche Nationalbibliothek lists this publication in the Deutsche Nationalbibliografie; detailed bibliographic data is available in the internet at http://dnb.d-nb.de.

Library of Congress Cataloging-in-Publication Data
Economic development in Central Asia : institutional underpinnings of factor markets / Joachim Ahrens, Herman W. Hoen (eds.).
 pages cm. — (Emerging markets studies ; vol. 6)
 ISBN 978-3-631-61215-6
 1. Economic development—Asia, Central. 2. Institutional economics. 3. Institution building—Asia, Central. 4. Asia, Central—Economic policy. 5. Asia, Central—Economic conditions. I. Ahrens, Joachim, 1963- II. Hoen, Herman Willem, 1960-
 HC420.3.E267 2014
 338.958—dc23

2014028267

ISSN 2190-099X
ISBN 978-3-631-61215-6 (Print)
E-ISBN 978-3-653-03226-0 (E-Book)
DOI 10.3726/978-3-653-03226-0

© Peter Lang GmbH
International Academic Publishers
Frankfurt am Main 2014
All rights reserved.
PL Academic Research is an Imprint of Peter Lang GmbH.

Peter Lang – Frankfurt am Main · Bern · Bruxelles · New York · Oxford · Warszawa · Wien

All parts of this publication are protected by copyright. Any utilisation outside the strict limits of the copyright law, without the permission of the publisher, is forbidden and liable to prosecution. This applies in particular to reproductions, translations, microfilming, and storage and processing in electronic retrieval systems.

This publication has been peer reviewed.

www.peterlang.com

Contents

Authors   *vii*

PART I: THEORETICAL AND CONCEPTUAL FOUNDATION   1

Joachim Ahrens and Herman W. Hoen
Institutional Change and Performance of Production
Factor Markets in Central Asia   3

Christian Danne
Commitment Devices, Opportunity Windows, and Institution
Building in Central Asia   17

PART II: CASE STUDIES   35

Saule Sagandykova
The Effects of Foreign Direct Investment on Wages in Kazakhstan   37

Abdul Ghaffar Mughal
Migration and Remittance Flows within and from Selected Countries
of Central Asia and the South Caucasus   59

Munira Aminova and Marc Jegers
Are Hofstede's National Cultures Homogeneous?
The Case of Uzbekistan   85

Kamiljon T. Akramov and Nurbek Omuraliev
Institutional Change and Agricultural Performance in Kyrgyzstan   99

Olav Heidelbach and Raushan Bokusheva
Factors Determining Crop Insurance Market Development in a Transition
Economy: The Case of Kazakhstan   127

Authors

Joachim Ahrens, Professor of International Economics, PFH Göttingen Business School, Germany

Kamiljon T. Akramov, Research Fellow at the International Food Policy Research Institute, Washington, DC, USA

Munira Aminova, Assistant Professor at Vesalius College, Vrije Universiteit Brussel, Belgium

Raushan Bokusheva, Senior Researcher, ETH Zurich (Swiss Federal Institute of Technology), Institute for Environmental Decisions, Zurich, Switzerland

Christian Danne, Macroeconomist, Trinity College Dublin, Ireland

Olav Heidelbach, European Union Programme Manager, Baku, Azerbaijan; previously Bishkek, Kyrgyzstan

Herman W. Hoen, Professor of International Political Economy, University of Groningen, The Netherlands

Marc Jegers, Professor at the Institute for European Studies, Vrije Universiteit Brussel, Belgium

Abdul Ghaffar Mughal, International Development Consultant, Los Angeles, CA, USA

Nurbek Omuraliev, Director of the Center for Science Methodology and Social Research, Academy of Sciences of the Kyrgyz Republic, Bishkek, Kyrgyzstan

Saule Sagandykova, Associate Professor at Kazakh-British Technical University, Almaty, Kazakhstan

PART I: Theoretical and Conceptual Foundation

Institutional Change and Performance of Production Factor Markets in Central Asia

Joachim Ahrens and Herman W. Hoen

1. INTRODUCTION

This book focuses on the economies in Central Asia including Kazakhstan, Kyrgyzstan, Tajikistan, Turkmenistan, and Uzbekistan. All countries have a communist past and, therefore, to a greater or lesser extent, still struggle with the legacy of mandatory Soviet-planning and large-scale state interference. Furthermore, they share other important characteristics. From a politico-economic perspective the following common features are relevant and need to be taken into account.

Firstly, Central Asia is exceptionally well-endowed with natural resources. In particular the proven stocks of oil and gas make the region distinct from other regions with a legacy of central planning, and it is obvious that these rich endowments did influence the chosen transition policies in the respective countries (Ahrens and Hoen 2013, 6). Moreover, the resource endowments triggered an increase in the importance of the geo-political position of the region.[1] It is beyond any doubt that, after the demise of the Soviet Union, Central Asia gained political influence in the world due to the exports of gas and oil (Pomfret 2006).

Secondly, the Central Asian countries are all landlocked. On the one hand, this implies that they are remote from important international markets and, therefore, are in a disadvantageous position. On the other hand, the geographically isolated position may as well appear as an incentive to create opportunities for becoming important transit countries, especially with the network of pipelines for oil and gas (Spechler and Spechler 2013, 218–9). In line with the earlier mentioned argument on endowments, this grid creates geo-political power of the region and, therefore, has security implications as well.

Thirdly, the transition to a market economy in Central Asia did not coincide with deliberate attempts to create Western type democracies. Whereas democratization fell short in many of the transition countries after the collapse of communism (Jeffries 2004), the creation of a market economy embedded in a

---

1 Many have coined this development as the 'New Great Game', thereby referring to the rivalry in the nineteenth century between the British and the Russian empires in Central Asia (Pomfret 1995, 21).

democratic order was not even intended to be established in Central Asia (Jeffries 2003). Kyrgyzstan may be conceived of as a slightly deviant case, but even in this most liberal country in the region autocracy and patrimonialism seem to be the rule (Ramas 2013).

In the fields of economics and political economy, scholarly interest in the region has been rather weak for a long period of time, or at best lagging behind the research focusing on transition in Eastern Europe and the European successor states of the Soviet Union.[2] From a politico-economic point of view, there are, however, good reasons to shift focus to the region of Central Asia. Firstly, the economic performance of the Central Asian countries has been surprisingly strong in relative terms. Secondly, the type of market economy that is emerging in the region does not resemble the institutional setting of 'Western style' capitalism, characterized with a limited role of the government that sets and monitors market rules, but rather reveals the development of a new type of state capitalism (Ahrens and Toews 2013). Both factors are the starting point of this book's topic and, therefore, need to be further concisely explored at the outset.

At the start of the transition to a market economy after the dissolution of the Soviet Union in 1991, the Central Asian countries were among the poorest successor states. Tajikistan's aggregate output per capita was less than half of the Soviet Union's total per capita output (46%). Uzbekistan (51%) and Kyrgyzstan (56%) were only slightly better off, whereas Turkmenistan (67%) and Kazakhstan (72%) were as well lagging far behind the poorest European republics, such as Azerbaijan (85%) and Moldova (85%) (van Selm 1995, 53).

In the 1990s, all five Central Asian countries were confronted with substantial declines in Gross Domestic Production (GDP). Kazakhstan was the first country that faced severe drops in production, but in comparison to Kyrgyzstan and Tajikistan the decline was still relatively modest. In the first half of the 1990s, Kyrgyzstan's GDP growth rates were -17% and even worse for a number of subsequent years.[3] Tajikistan suffered from a civil war (1992–1997), which, of course, had a devastating impact on its economy. Turkmenistan followed a similar path as Kazakhstan, be it that the latter country's decline and recovery followed with a time lag. In the first years of transition, Uzbekistan was by far

---

2 Seminal books on the transition to a market economy, such as van Brabant (1998), Gross and Steinherr (1995), and Roland (2000), are only marginally paying attention to Central Asia. So does Åslund (2002), but in Åslund (2007), Central Asian transition processes are included.
3 GDP figures are all taken from the EBRD, *Transition Report*, various years, unless otherwise stated.

the best performing country in Central Asia, which took many scholars by surprise (Zettelmeyer 1999).

At the end of the 1990s, the Central Asian countries began to recover from the downturn. In terms of GDP per capita, Turkmenistan has been the most successful economy. At the turn of the millennium, it had reached its GDP level of 1989 and by the end of 2012 this well-endowed country had been able to quadruple the size of its economy. Notably gas exports led to a GDP that was nearly four times its level of 1989.[4] Uzbekistan was in second position. It had more than doubled its economy since 1989. This was predominantly due to cotton exports that could quickly be shifted away from Russia to world markets (Hoen 2013a). Kazakhstan had a similar overall performance as Uzbekistan, albeit its growth rates were a little more moderate, which led to a GDP level in 2012 of 180% of that in 1989. Whereas Turkmenistan, Uzbekistan and Kazakhstan have been able to perform better than the average of all transition countries, Kyrgyzstan and Tajikistan have not. The turnaround in these mountainous countries was far less convincing than in the neighbouring countries. In 2012, both countries had reached their GDP levels of 1989, but they have been relatively underperforming.[5]

The economic performance of the Central Asian countries since their independence may have been successful in relative terms, but the emerging type of capitalism is even more surprising and remarkable. Distinct varieties of state capitalism are emerging. State capitalism is rather common in large parts of Asia (Bremmer 2009). It may embrace globalization, funding from international financial institutions, popular welfare scientific management, small businesses, and even listing minority shares on the stock exchanges of the world. But state capitalism does *not* allow investors to pick winners and losers for investing; does not allow its currency to fluctuate wildly; does not allow off-shoring and tax avoidance; and does not make indelible commitments to multinational corporations or international banks.[6]

---

4   Turkmenistan's gas industry accounts for half of its GDP (Repkine 2004, 160). Despite the statistical reliability of the data published (Pomfret 2006, 107–120), it is beyond any doubt that Turkmenistan has been the best-performing country in the region.
5   In 2012, the average GDP of all transition countries was 160% (1989 = 100).
6   Authoritarian governments like those of follower countries like Meiji Japan, Park Chung Hee's South Korea, and Lee Kwan Yew's Singapore successfully adopted 'state capitalism' with 'Asian values' during several decades following a major break in their national histories. They created 'national champions,' '*chaebols*', or 'strategic industries'. Legacies of epochs of 'state capitalism' still exist. Those include huge state-owned companies like Japan Post Holdings, Daewoo, and Singapore Airlines (Economist 2012; Stark 2012).

The incomplete transitions to developed market economies in Central Asia have been characterized by the emergence of state capitalist economies with two new sectors. A dual 'state capitalism' seems to be surfacing. This consists of a core sector, benefiting from the countries' natural endowments and exports to global markets. Material support of the independent authoritarian governments is being financed by their 'staple globalism' strategy, which involves state direction of exports and select imports. Alongside the core sector is a growing periphery of service and small business enterprises operating under market conditions with some state intervention. Whereas in the core, the state is picking the winners, in the periphery the market is (Hoen 2013b, 17–8).

The relatively strong economic performance since the collapse of communism and the distinct type of a capitalist mode that is emerging in Central Asian countries, triggers many thought-provoking research questions on institution building and institutional performance. To what extent is the emergence of a market economy in Central Asia significantly distinct from developments in other transition countries? What are the building blocks of the Central Asian variety of capitalism? What factors explain the emergence of a distinct variety of capitalism in Central Asia? To what extent does a new variety of capitalism in Central Asia enlighten the good performance of their economies?

The research focus of the book is a cross-section of the questions raised above in which factor markets and institutions, as fundamental drivers of economic growth, serve as the common denominator. The volume scrutinizes the design and performance of factor markets in Central Asia: (i) land and natural resources, (ii) labour, and (iii) capital. The last-mentioned category knowingly encompasses entrepreneurship, but there are also sound arguments to position it as a category of its own.

In an attempt to underpin the theoretical strands of the book, the remainder of this chapter is structured as followed. Section 2 elaborates on what kind of research currently is conducted in the field of institutional economics and Central Asia. In other words, it raises the question: 'Where do we stand?' Subsequently, Section 3 zooms in on this book's particular focus, which is the institutional change of production factor markets in Central Asia. Factor markets are distinct. The transition after the collapse of communism in Central and Eastern Europe has revealed that the emergence of fully-fledged markets is a strenuous process with far-reaching implications for all sectors in the economy (van Brabant 1998). But, relatively speaking, consumer goods markets seem to have adjusted relatively quickly – painful as it may have been. Factor markers, however, have not (Blanchard 1997). Section 3, therefore, scrutinizes the impact for the transition process. Section 4 concludes with an overview of the structure of the book and an outline of the remaining chapters.

2. INSTITUTIONAL ECONOMIC RESEARCH ON CENTRAL ASIA: WHERE DOES IT STAND?

Institutional economics is often seen as a kind of a black box. It is widely accepted that institutions matter, but how they can best be designed, in what way they need to be and how they develop over time is far from clear (Wagener 1993). In the initial stages of transition, economics has largely neglected societal conditions and political interests. Now it has become clear that these have to be taken into account for achieving politically feasible economic transition (Ahrens and Hoen 2013, 9).

In the so-called second-generation reforms the notion was taken into account, but as was already stated above, Central Asia has not very much been in the focus of institutional economists and political scientists. Recently, however, there is an upsurge of research conducted in the field. Considering institution building in and institutional performance of the transition countries in Central Asia, three building blocks can be conceived of in which energy forms the leading thread. Firstly, there is a vastly growing amount of research on the economic performance of Central Asia during transition. Economists have taken the lead in this research. Secondly, due to the rich endowments a lot of political economic research has been conducted on the question of the resource curse and more specifically on obstacles to democratization and the emergence of a *rentier* state in Central Asia. Thirdly, based upon the natural resources, political scientists and international relations scholars have focused research on Central Asia and the geopolitics of energy. The three building blocks of research directly follow from the characteristics indicated in Section 1. In the current section, they serve as a guideline for examining where the discipline of institutional economics stands with respect to research conducted on Central Asia. Subsequently, the revealing 'insufficiencies' give rise to shift the research focus in Section 3 and to underpin the need for studying factor markets from an institutional economics research perspective.

2.1 Economic performance of Central Asia during transition

Although the literature on the economic performance of Central Asia is not that massive, there is substantial research on macro-economic developments since independence and on the micro-economic underpinnings of transition (EBRD various years; Mogilevski and Hasanov 2004; Pomfret 1995; 2006; Repkine 2004; Spechler 2008; Spechler *et al.* 2014; Zettelmeyer 1999). The literature is predominantly macro-economic in nature and focuses on economic growth, exchange rate regimes, unemployment, inflation, and income distribution. Kazakhstan and Uzbekistan stand out as examples of resource driven (and in the Uzbek case somewhat unexpected) growth. Kyrgyzstan has been discussed as a liberal reformer that failed to create sustainable growth, whereas Tajikistan represents the tragic case of a country stuck in violence and poverty. Turkmenistan's growth performance is either distrusted or attributed to the exports of gas.

In the discussion of the micro-economic underpinnings of transition, the literature centres on efficiency gains. First and foremost, the privatization process has been under scrutiny (Åslund 2007; Pomfret 2006), but there is also research conducted on finance and banking and the extent to which the monetary sector acts independently from politics (EBRD, various years). It is safe to conclude that micro-economic research focuses on the conditions which have to be met in order to increase efficiency. It is less outspoken in its analysis on what happens if these conditions are not fulfilled, let alone in exploring the factors that explain why efficiency gains did not materialize. In particular the latter would imply a need to focus on the elucidation of institutional change. The implementation of market rules or lack thereof is not something that can be 'mechanically' engineered, but is the result of dissimilar economic ideas and interests. The research on these micro-economic underpinnings of transition is to a large extent lacking for Central Asia.[7]

2.2 Natural endowments, obstacles to democratization, and the rentier state in Central Asia

Economic performance cannot be disentangled from nature's endowments in Central Asia. Turkmenistan has the largest proven gas reserves in the world (Cooley 2012), and it is beyond any doubt that the economic performance of Kazakhstan is resource driven (Akram et al. 2006). To a lesser extent, but nevertheless significantly, gas exploitation fuels Uzbekistan's economy (Repkine 2004). The rich endowments of natural resources, notably oil and gas, have been an incentive for scholars to take stock of the theory of the resource curse and to see to what extent it can deliver valuable insights for the state-driven nature of the countries in Central Asia.

The theory of the resource curse has many fathers and this section attempts to pursue only limited genealogical research. The basic premise of the theory is straightforward. Countries with an abundance – albeit finite amount – of non-renewable natural resources tend to have less economic growth and experience

---

7   See, however, Ahrens and Hoen (2013) for an overview.

inferior development outcomes than countries that are not particularly well-endowed.[8] Following a straightforward economic line of reasoning, which also perfectly fits as well under the heading of the former sub-section, there is the concept of the *'Dutch disease'*. This theory boils down to the problem of a real appreciation of the currency due to energy resources.[9]

There are, however, also explanations that rather focus on the domestic side of the economy. In this context, corruption and the close ties between politicians and businessmen stand out. It is argued that in resource-abundant states politicians tend to stay longer in power through allocating resources to supportive elites rather than by investing in the diversification and modernization of the economy (Collins 2006). This is not just a matter of conflicting interests. Diversifying interests are cleared in a fraudulent manner. In an attempt to keep support, politicians and bureaucrats give all kinds of preferential treatments, either in the form of bribes or in any other kind of tax exempts, loans, privileged price guarantees for demand, *et cetera*. These are all benefits that others are refrained from, because they are not seen as part of the distinct political elites. It is important to indicate, though, that there is not always a clear line to be drawn with which a distinction can be made between *just* and *illegitimate* (Cooley 2012, 138). Resource curse research on Central Asia encompasses those studies that focus on the "political outcomes of the merger between businesses and bureaucracies" (Rustemova 2011, 31). This specific merger of the political and the economic, or, for that matter, the state and the market, is also coined as the *rentier state* (Mahdavy 1970).

In the case of Kazakhstan, there are hard empirical findings that indeed ground the concept of the *rentier state*. Research reveals that it is beyond the slightest doubt that Kazakhstan's growth is fueled by its natural resources, notably oil and to a lesser extent gas. So, the production structure is pivotal. Kazakhstan's rents from oil are 27.2% of GDP, whereas gas rents amount to an estimated

---

8 Ample of empirical evidence has been presented and much attention has been given to persistently underperforming economies in Latin America and Africa (Sachs and Warner 2001). Since resource-dependency is often measured as primary exports as percentage of the gross domestic product, it may come as no surprise that the explanations for the observed underperformance of well-endowed countries often focus on the vulnerability of export earnings, the crowding out of the manufacturing sector, and the inclination of excessive state borrowing due to optimistic expectation of future earnings (Sali-i-Martin and Subramanian 2003).
9 The currency appreciation has a negative impact on competitiveness of the manufacturing sector. As a result of the crowding out, economic performance will decline. For the theory of the *Dutch disease* the reader is, e.g., referred to Cordon (1984) and van Wijnbergen (1984). The application of the theory to the countries in Central Asia is conducted by Ross (2009).

1.5% (Akram *et al.* 2006, 43)). The merger between the state bureaucracy and the energy sector seems to be undisputed as well, which in the literature also has been coined "resource nationalism", i.e. "policies that are intended to increase state control over natural resources vis-à-vis private actors" (Nurmakov 2010, 21). The Uzbek case does not as easily fit in the model of the resource curse. Of course, the country has oil and gas rents as, but there seem to be agricultural rents from cotton. It has been pointed out though that cotton rents are much smaller for the state, because the production of cotton involves much more actors than is the case for mineral fuels. So, total rents do not flow to the state, but are distributed over many others (Rustemova 2011, 33).

2.3 Central Asia and the geopolitics of energy

A third theme that can be discerned in research conducted on the political economy of Central Asia is the rise of the region's geopolitical influence. Whereas studies on the '*rentier state*' focus on domestic relations, the geopolitical literature pays specific attention to the external dimension. The leading thread, of course, is the same: energy.[10] Considering regional economic power relations, academic interest reveals, on the one hand, a trade-off between Russia and China, and, on the other hand, attempts of the European Union (EU) in cooperation with the European Bank for Reconstruction and Development (EBRD) to fulfil its neighbourhood policy commitments. Since the institutional economics underpinnings of geopolitics in Central Asia are rather weak, both aspects will only briefly be touched upon.

Gas and oil supplies have drastically changed power relations in the region. The delicate relationship between Kazakhstan and Russia has been extensively studied (Sievers 2003), but the trade-off between regional hegemonic positions of Russia and China is a new focus. Oil and gas deliveries require large-scale infrastructural investments and, to an increasing extent, the grids are financed with Chinese capital (Spechler and Spechler 2013).

Considering the theoretical and empirical underpinnings of external help, there is a stream of literature about policy transfers from the EU to Central Asia in the Partnership and Cooperation Agreement (De Deugd 2013; Warkotsch 2011). The EU neighbourhood policies zooms first and foremost in – again – on energy (Götz 2011), but the dialogue expands in the fields of environmental protection (Partzsch 2007) and, lifting microeconomic barriers to gain market efficiency

---

10 There is, however, also a growing amount of economic literature that focuses on economic support to increase regional security. That holds in particular for Uzbekistan that facilitates American military operations in Afghanistan (Spechler 2008, 111; Spechler. and Spechler 2013).

(Schmitz 2007). From an institutional economics perspective the latter is highly interesting. However, that perspective is lacking in the scholarly contributions so far. All the relevant research sticks to a 'mechanical' way of reform tools that need to satisfy the neoclassical frame of welfare conditions rather than taking into account the institutional legacy and the dynamics of institutional modification over time.

3. INSTITUTIONAL CHANGE OF PRODUCTION FACTOR MARKETS IN CENTRAL ASIA

Section 2 has revealed that research conducted on transition processes in Central Asia emphasizes the impact of recourse endowments on either the domestic economy or the regional geopolitical situation. Energy matters. Hence, land and its endowments have been taken into account. There has been a vast and growing research field on this factor market in Central Asia. Nonetheless, the underlying assumption of this volume is that factor markets are not extensively studied from an institutional economic research perspective. This section seeks to clarify that position and explicates the need to fill the gap.

Institutional change is a complicated matter. It is clear that institutions are not only a result of deliberate design and implementation. Their existence and efficacy are also the outcome of organic evolutionary change (Wagener 1993). Taking the legacy of the past – history in general – into account implies that institutional change is path dependent and irreversible. The longer the time span for institutionally creating fully-fledged markets, the higher is the probability of path dependency.

Empirical evidence has shown that the process of developing functioning consumer goods markets takes time. For factor markets – land, labour, and capital – the time span is considerably longer for a number of reasons. Firstly, they rely on long-term investments, whether that is educational or vocational training of labourers, long-term capital investments. Secondly, the stickiness of behaviour of the relevant actors, in particular the ones in labour markets and entrepreneurship, requires more time. Thirdly, factor markets respond to expected and effective demand. An increase in the demand for final products leads to a consecutive increase in the demand for productive resources.

The longer time span of maturing factor markets implies that it is more likely that the emerging institutional underpinnings of the markets deviate from initially designed plans. Moreover, Central Asian factor markets will not only be distinct from the original ideas, they will also deviate from the factor markets which developed in other transition countries. One cannot expect the emergence of these markets to lag behind the development of factor markets in more advanced transition economies. Since market development is greatly shaped by the country-specific institutional environment and government policy, factor market developments in Central Asia will show significant idiosyncratic characteristics.

One of the most important aspects of Central Asian transition is the role of the state. As was explained above, the state picks the winners and, as such, is directly interfering in factor markets. In many instances, it operates as an entrepreneur. This peculiar mode of capitalism, which has been coined *state capitalism*, appears to stimulate the emergence of a relatively well-performing market-based economy, and one of the crucial questions is to what extent institutional design and performance of factor markets in Central Asia are distinct and how the discrepancy can be explained.

This book contributes to a stock taking of the emergence of factor markets in Central Asia and seeks to find explanations from an institutional economics perspective. It is genuinely believed that this research fills a gap in the transition literature and elucidates the process of politico-economic development.

4. OUTLINE OF THE BOOK

This book was written within the framework of the project *Emerging Market Economies in Central Asia: The Role of Institutional Complementarities in Reform Processes*, which was encouraged, supported and financed by the *VolkswagenFoundation*.[11] It consists of two parts: The first part provides the theoretical and conceptual foundation and addresses the emergence of economic institutions in Central Asia. Part II discusses case studies on institutional performance in different factor markets.

Part I consists of two chapters. Besides this introductory chapter, it includes Christian Danne's contribution on commitment devices and windows of opportunities in the process of institution building in Central Asia. Its basic proposition is that the lack of appropriate incentives explains the lack of reforms in Central Asia. That argument is underpinned by focusing on the education system.

Part II presents a number of case studies on the institutional performance of Central Asian countries in the field of factor markets. The case studies vary in terms of countries as well as in the modes of the factor market. Chapters 3 and 4

---

11 The editors are grateful to Joachim Algermissen, Nico Stobinsky, Nikolas Weise, Franziska Wölbern, Katharina Wölbern for research and editorial assistance.

focus on the capital and the labour market, respectively. In Chapter 3, Saule Sagandykova analyses the impact of foreign direct investment on wages in Kazakhstan. Chapter 4, written by Abdul Ghaffar Mughal, scrutinizes migration and remittances flows in Central Asia and conducts the analysis by using the Caucasus as a point of reference. Subsequently, Chapter 5 by Munira Aminova and Marc Jegers explore the importance of informal institutions and more specifically cultural aspects by applying Hofstede's approach. They assert that intra-country cultural heterogeneity is enormous in Central Asia and hence may have distinct effects on economic performance which are hard to predict.

Part II concludes with two chapters which focus on land as one of the factor markets. Chapter 6 deals with institutional reform and agricultural performance in Kyrgyzstan. Kamiljon T. Akramov and Nurbek Omuraliev pinpoint the complete liquidation of large-scale state and collective farms in which Kyrgyzstan was quite unique. Olav Heidelbach and Raushan Bokusheva conclude the book with a chapter on the development of crop insurance markets in Kazakhstan. The authors reveal the weaknesses to set up an appropriate incentive system and underline the importance of such a system for transition countries.

In sum, institution building in transition countries has been an arduous process. That holds for all transition countries. Central Asia is no exception. On the contrary, the region reveals patterns of institutional change that are significantly distinct. It is often stated that globalization makes a 'smaller' world by institutional convergence. This book challenges that conviction. Whereas in the first two decades following the demise of communism the scholarly debates predominantly focused on the implementation of an allegedly known economic order – a market economy – the emergence of diverging market economies needs now more decidedly taken into account. This book does exactly that. Focussing on the institutional design and performance of factor markets, it shifts the focus on emerging markets in Central Asia from an instrumental one about the "means" to one about the "ends" (Bohle and Greskovits 2012, 6).

REFERENCES

Ahrens, J. and H.W. Hoen (2013), 'Economic Transition in institutional change in Central Asia,' in J. Ahrens and H.W. Hoen (eds), *Institutional Reform in Central Asia. Politico-economic challenges*, London and New York: Routledge: 3–17.

Ahrens, J. and G. Toews (2013), *Varieties of Capitalism in Developed and Transition Countries: an Empirical Investigation of Economic Convergence of OECD and Post-Soviet Countries*, Unpublished Paper, Göttingen: PFH.

Akram, E., M. Raiser and W.H. Buiter (2006), 'Nature's Blessing or Nature's curse? The Political Economy of Transition in Resource-based Economies,' in R.M. Auty, I. de Soysa (eds), *Energy, Wealth and Governance in the Caucasus and central Asia*, New York: Routledge: 39–56.

Åslund, A. (2002), *Building Capitalism. The Transformation of the Former Soviet Bloc*, Cambridge: Cambridge University Press.

Åslund, A. (2007), *How Capitalism was Built. The Transformation of Central and Eastern Europe, Russia, and Central Asia*, Cambridge: Cambridge University Press.

Blanchard, O. (1997), *The Economics of Post-communist Transition*, Oxford: Clarendon Press.

Bohle, D. and B. Greskovits (2012), *Capitalist Diversity on Europe's Periphery*. Ithaca, NY: Cornell University Press.

Brabant, J.M. van (1998), *The Political Economy of Transition. Coming to grips with history and methodology*, New York and London: Routledge.

Bremmer, I. (2009), 'State Capitalism comes of age. The End of the Free Market?' *Foreign Affairs*, 88: 40–55.

Collins, K.A. (2006), *Clan Politics and Regime Transition in Central Asia*. New York, NY: Cambridge University Press.

Cordon, W.M. (1984), 'Booming Sector and Dutch Disease Economics: Survey and Consolidation,' *Oxford Economic Papers*, 36: 359–380.

Deugd, N. de (2013), 'Policy Transfer between the European Union and the Countries from Central Asia,' in J. Ahrens and H.W. Hoen (eds), *Institutional Reform in Central Asia. Politico-economic challenges*, London and New York, NY: Routledge: 232–247.

EBRD (European Bank for Reconstruction and Development) (various years), *Transition Report*, London: EBRD.

The Economist (2012), 'Special Report – State Capitalism. The Visible Hand,' January 2012.

Götz, G. (2011), 'Energy cooperation: the Southern gas transport corridor,' in A. Warkotsch (ed.), *The European Union and Central Asia*, London and New York: Routledge: 148–162.

Gross, D. and A. Steinherr (1995), *Winds of Change. Economic Transition in Central and Eastern Europe*, London and New York: Longman.

Hoen, H.W. (2013a), 'De Economie van Oezbekistan: Het Spel en de Knikkers,' *Prospekt* (http://www.prospekt-online.nl/prospekt/prospekt_ artikelen_2013/ knikkers.html).

Hoen, H.W. (2013b), 'Emerging Market Economies and the Financial Crisis: Is there Institutional Convergence between Europe and Asia?' *Discourses in Social Market Economy* 4: 1–27.

Jeffries, I. (2003), *The Caucasus and Central Asian Republics at the Turn of the Twenty-first Century: A Guide to the Economies in Transition*, London and New York: Routledge.

Jeffries, I. (2004), *The Countries of the Former Soviet Union at the Turn of the Twenty-first Century: The Baltic and European States in Transition*, London and New York: Routledge.

Mogilevsky, R. and R. Hasanov (2004), 'Economic Growth in Kyrgyzstan,' in G. Ofer and R. Pomfret (eds), *The Economic Prospect of the CIS. Sources of Long Term Growth*, Cheltenham: Edward Elgar: 224–248.

Partzsh, L. (2007), *Global Governance in Partnerschaft. Die EU-Initiatieve 'Water for Life'* Baden-Baden: Nomos.

Pomfret, R. (1995), *The Economies of Central Asia*. Princeton: Princeton University Press.

Pomfret, R. (2006), *The Central Asian Economies since Independence*, Princeton and Oxford: Princeton University Press.

Ramas, R.R. (2013), 'The Institutional Persistence of Patrimonialism in the Kyrgyz Republic: Testing a Path Dependency (1991–2010),' in J. Ahrens and H.W. Hoen (eds), *Institutional Reform in Central Asia. Politico-economic challenges*, London and New York: Routledge: 129–151.

Repkine, A. (2004), 'Turkmenistan: economic autocracy and recent growth performance,' in G. Ofer and R. Pomfret (eds), *The Economic Prospect of the CIS. Sources of Long Term Growth*, Cheltenham: Edward Elgar: 155–176.

Ross, E. (2009), 'Oil Funds for Oil Blessings. The Performance of Non-renewable Resource Funds in Combating the Resource Curse: The Cases of Azerbaijan and Kazakhstan,' Working Paper. Göttingen: PFH.

Rustemova, A. (2011), 'Political Economy of Central Asia: Initial Reflections on the Need for a New Approach,' *Journal of Eurasian Studies* 2: 30–39.

Sachs, J.D. and A.M. Warner (2001), 'Natural Resources and Economic Development. The Curse of Natural Resources,' *European Economic Review* 45: 827–838.

Sala-i-Martin, Xavier and Arvind Subramanian (2003), 'Addressing the National Resource Curse. An Illustration from Nigeria,' *National Bureau of Economic Research*, Woking Paper 9804, Cambridge, MA: NBER.

Schmitz, A. (2008), 'Efficiency and its costs: The "Strategy for a New Partnership" with Central Asia,' in D. Kietz and V. Perthes (eds), The Potential of the Council Presidency. An Analysis of Germany's Chairmanship of the EU, 2007', *Stiftung Wissenschaft und Politik*, SWP Research Paper RP01: Berlin.

Selm, B. van (1995), *The Economics of Soviet Break-up*, Capelle a/d IJssel: Labyrint Publications.

Sievers, E.W. (2003), *The Post-Soviet Decline of Central Asia. Sustainable Development and Comprehensive Capital*, London: RoutledgeCurzon.

Spechler, M.C. (2008), *The Political Economy of Reform in Central Asia. Uzbekistan under Authoritarianism*, London and New York: Routledge.

Spechler, M.C., K. Bektemirov, S. Chepel and F. Suvankulov (2004), 'The Uzbek Paradox: Progress without Neo-liberal Reform,' in G. Ofer and R. Pomfret (eds), *The Economic Prospect of the CIS. Sources of Long Term Growth*, Cheltenham: Edward Elgar: 177–197.

Spechler, D.R. and M.C. Spechler (2013), 'The USA and Central Asia: intermittent allies,' in J. Ahrens and H.W. Hoen (eds), *Institutional Reform in Central Asia. Politico-economic challenges*, London and New York: Routledge: 248–256.

Spechler, M.C. and D.R. Spechler (2013), 'Will Russia regain its dominant role in Central Asia?' In J. Ahrens and H.W. Hoen (eds), *Institutional Reform in Central Asia. Politico-economic challenges*, London and New York: Routledge: 215–227.

Stark, M. (2012), *The Emergence of Developmental States from a New Institutionalist Perspective. A Comparative Analysis of East Asia and Central Asia*, Frankfurt: Peter Lang.

Wagener, H.-J. (1993), 'Some Theory of Systemic Change and Transformation,' in H.-J. Wagener (ed.), *On the Theory and Policy of Systemic Change*, Heidelberg and New York: Physica-Verlag: 1–20.

Warkotsch, A. (2011), 'The EU and Central Asian geopolitics,' in A. Warkotsch (ed.), *The European Union and Central Asia*, London and New York: Routledge: 63–73.

Wijnbergen, S. van (1984), 'Dutch Disease: A Disease after All?' *The Economic Journal* 94: 41–55.

Zettelmeyer, J. (1999), 'The Uzbek Growth Puzzle,' *IMF Staff Papers* 46(3): 274–292.

Commitment Devices, Opportunity Windows, and Institution Building in Central Asia[*]

Christian Danne[†]

1. INTRODUCTION

Sound economic institutions have been found to be an essential factor in developing a country's wealth and long-run growth. Reliable institutions reduce uncertainty and transaction costs and thereby foster investment and economic growth (Hall and Jones 1999; Acemoglu *et al.* 2005).

This paper analyses the institutional reform process in Central Asia from 1995 to 2006. I compare the reform processes in Central Asian economies with those conducted in emerging market economies in Central and (South) Eastern European countries (CEEC), and the Middle Eastern and North African (MENA) economies, in order to identify factors that have helped Central Asian countries with their transition process and institutional reforms.[1][2] CEEC and MENA economies share several similarities, as well as important differences, in terms of history, economic structure and culture. This set-up provides a natural experiment that allows us to identify potential drivers of the reform process. First, each of these countries can be considered as emerging markets with a large fraction of external trade, whereas CEEC and Central Asian economies are former socialist states, while the MENA economies are mostly former colonies of European countries. Second, both Central Asian and MENA economies have large shares of commodity exports, as opposed to the resource scarce CEEC, which have a large fraction of commodity imports. Third, all three regions have cooperation agreements with the European Union (EU) at varying levels of intensity, ranging from EU Accession Country status to very loose agreements regarding technical assistance and economic and cultural exchange (such as

---

[*] I am grateful to Svetlana Bratakova, Michael Breen, Benjamin Elsner, Michael O'Grady, Andreas Tudyka, Julia Matz, and the participants of the conference on "Institutions, Institutional Change, and Economic Performance in Central Asia" in Goettingen in September 2008 for their comments and suggestions. I would also like to thank Jan Teorell, Nicholas Charron, Marcus Samanni, Soren Holmberg and Bo Rothstein for making their Quality of Government Dataset 2009 available at http://www.qog.pol.gu.se.

[†] Email: dannec@tcd.ie. Funding from the Irish Research Council for the Humanities and Social Sciences (IRCHSS) is gratefully acknowledged.

[12] The terms CEEC and MENA are used very loosely in this study. The group of CEEC countries includes all Central and (South) Eastern European countries and the Baltic States that had a (potential) EU candidate status in 2006. The group of the MENA economies is restricted to MENA countries that are part of the Barcelona Process. For a detailed description of the country groups see the Appendix.

Technical Assistance for the Commonwealth of Independent States (TACIS) programme).

There is a vast literature on the determinants of economic institutions and political change. Most papers focus on historical and exogenous factors, such as the abundance of natural resources or colonial and legal origins (Peters 1996; Acemoglu *et al.* 2001; Beck and Laeven 2006). This paper argues that institutional arrangements, first and foremost, are subject to a political decision-making process in which politicians are responsible for setting-up institutional arrangements, regardless of the actual political system. Exogenous factors, by definition, cannot be changed in order to achieve better institutional settings and thus, offer little for policy recommendations. Looking at institutions from a choice perspective allows for the identification of factors that can be subject to change and can therefore lead to institutional reforms.

In short, the paper looks at two types of factors. First, the paper identifies factors that may have hindered institutional reforms in Central Asia and argues that the lack of reforms in Central Asia is the result of a lack of incentives for policy makers and individuals. Second, the paper provides stylised evidence on endogenous external factors that have helped Central Asian economies, in the past decade, to overcome reform inertia and revise existing institutional arrangements. Based on the findings, it is argued that deficiencies in the education system and preferences about institutional reforms are an important factor for the persistence of institutional settings in Central Asia. Preferences and the education received under the old socialistic system are two important channels through which historical factors affect current institutional arrangements and provide an obstacle for implementing more efficient, market-based institutions. Second, external factors such as real and financial openness, combined with fixed exchange rates, have constrained policy making towards better institutional arrangements. A high degree of trade and financial openness provides an incentive to reform institutional settings, as countries have a larger share of revenues per gross domestic product (GDP) from external trade and a larger share of foreign investment in total investment. Therefore, opportunity costs in terms of forgone business opportunities as a result of bad governance provide an incentive to improve the institutional set-up. Third, financial openness provides an incentive to remedy deficiencies in regulatory frameworks, thereby making the economy less prone to sudden capital outflows which may result in financial turmoil and painful economic recessions. Pegging the exchange rate, in combination with financial openness, can have the same effect and amplifies the effect of financial openness, as undesirable domestic policy measures may result in speculative attacks against the currency regime. Forth, the TACIS-programme has likely contributed positively as well, as it explicitly aims for institutional change and more market-oriented institutions which are essential for closer ties with the EU and better access to EU goods and financial markets.

Lastly, the role of economic shocks, such as a domestic economic crisis, may speed up the process of institutional change as it may shift preferences of politicians and individuals towards a better institutional framework.

The remainder of the paper is organized as follows. Section 2 discusses the sources of persistence of institutional arrangements. Section 3 discusses the external factors that have helped Central Asian countries to overcome persistence in the reform process and the mechanisms behind them. Section 4 concludes.

2. PERSISTANCE OF INSTITUTIONAL ARRANGEMENTS

Since the breakdown of the socialist systems in the early 1990s, transition economies in Central Asia have faced the problem of rearranging their institutional framework and finding strategies to steer economic growth in order to smoothen the structural adjustment process towards a market economy.

Figure 1 compares the quality of economic institutions in Central Asian economies against an unweighted average of the EU-15, CEEC, MENA economies, and Russia as benchmarks using the Economic Freedom Index provided by the Heritage Foundation (2008).[13] The index is a *de facto* measure of institutional quality and defines institutions as mechanisms to ensure property rights and efficient public bodies, in order to provide public goods in an efficient way and reduce transaction costs. Institutional quality is measured using 10 different sub-categories, namely business and investment regulations; trade and financial sector regulations; monetary and fiscal institutions; property rights protection and corruption; and labor market institutions. Each of the 10 categories is graded on a continuous scale from 0 to 100, with 100 representing a minimum of distortions and costs associated with the existing arrangements and 0 a maximum. Figure 1 reveals that the institutional reform process in Central Asia, CEEC and the MENA economies has been heterogeneous. Although Figure 1 suggests that institutional settings in all countries are quite

---

13 Throughout the paper unweighted averages are used for the CEEC, MENA, and EU-15 countries in the graphs in order to compare policies without distorting size effects of countries. While, the primary focus is on the comparison between the small open economies in the three regions, Russia and the EU-15 are added to the analysis as both countries represent important hegemonic states with quite opposing institutional set-ups that exert considerable influence on Central Asian economies.

persistent, the transition economies in CEEC and, lately, Central Asia appear to be relatively successful in reforming their institutional arrangements. While Central Asian economies started from a low level in the early and mid-1990s, they have caught up with their neighbors in the MENA region in recent years, despite poorer starting conditions. Some Central Asian countries, namely Kazakhstan and the Kyrgyz Republic, have even come close to CEEC levels, even though CEEC started off at a much higher level of institutional quality in the mid-1990s. Particularly, the Kyrgyz Republic, Tajikistan, and Kazakhstan show a significant upward trend after the Russian debt crisis in 2001 and became the most dynamic reformers in the group, while Turkmenistan and Uzbekistan are still lagging behind. Leaving aside Turkmenistan, all Central Asian countries show some co-movements with Russia and their surrounding countries in Central and Eastern Europe.[14] This may suggest that there are factors that are common to either Central Asia and Russia, or Central Asia and CEEC, which may have driven the reform process in the region.

Figure 1: Institutional Quality 1995–2006

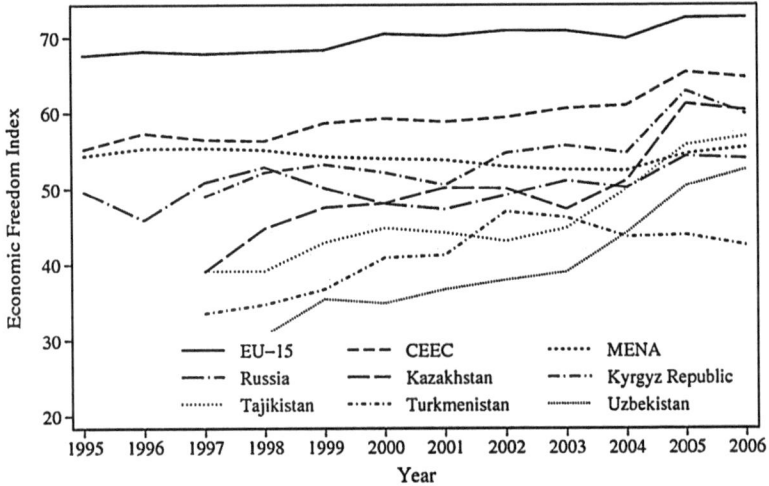

Note: No data available for Central Asian countries prior to 1997.
Source: Heritage Foundation (2008).

---

14 In recent years, China's influence on Central Asian countries has grown considerably through, e.g., the establishment of the Shanghai Cooperation Organisation in 1996 as well as through economic ties. For simplicity and historical reasons, Russia is chosen as a counter-benchmark to the EU. But in the future, the Chinese influence in the region should be monitored.

Although most Central Asian countries show signs of institutional change in recent years according to Figure 1, it is important to understand the factors that cause the persistence of lumpy institutions, in order to ensure that the reform process will be continued and not be reversed.

Several authors have discussed factors behind the persistence of institutional arrangements (International Monetary Fund (IMF) (2005) for an overview). The most prominent argument as to why bad policies persist is that the political elite has control over economic rents from natural resources, such as crude oil (Sachs and Warner 2001; Ross 2001; Sala-i-Martin and Subramanian 2003; Ramsay 2006). Access to these rents pose a disincentive for political and economic elites to reform institutions, as (tax) earnings from other economic sectors, which would need sound institutional frameworks and property rights protection for their development, pale in insignificance (Rajan and Zingales 2006; Congdon Fors and Olsson 2007). Moreover, poor property rights protection can be used to block the introduction of new technologies that may reduce the elites' future political and economic power (Rajan and Zingales 2006; Acemoglu and Robinson 2006).

Figure 2 shows the share of oil related GDP as a share of total GDP as a measure of resource dependence in Central Asia. Oil related GDP ratios in Central Asia are compared to an unweighted MENA average. It is apparent that Kazakhstan and the Kyrgyz Republic have the highest share of oil related GDP of any of the Central Asian states. All three countries are well above the average MENA country, whereas Uzbekistan is roughly at the same level and Turkmenistan and Tajikistan are below the MENA average.

Comparing this graph to Figure 1, it seems that no clear relationship between resource dependence and institutions exists in Central Asia. Kazakhstan and the Kyrgyz Republic are the countries with the highest fraction of oil related GDP as well as the best institutions among Central Asian economies. The countries with the poorest institutions in the Central Asian group are also the ones with the lowest degree of resource dependence according to our measure. While this does not necessarily reject the theory of a resource curse for Central Asia, it might indicate that other factors have mitigated the effects of the easy-rents sector on institutional settings.

Another argument proposed in the literature is that being a former colony or having a legal system that roots in the European legal tradition, has a positive long-term effect on institutional arrangements (Acemoglu *et al.* 2001; Kuran 2004; La Porta *et al.* 2008). Once modern, market-based institutions were implemented during colonial times, they have not been reversed after the end of

Figure 2: Oil related GDP 1995–2006

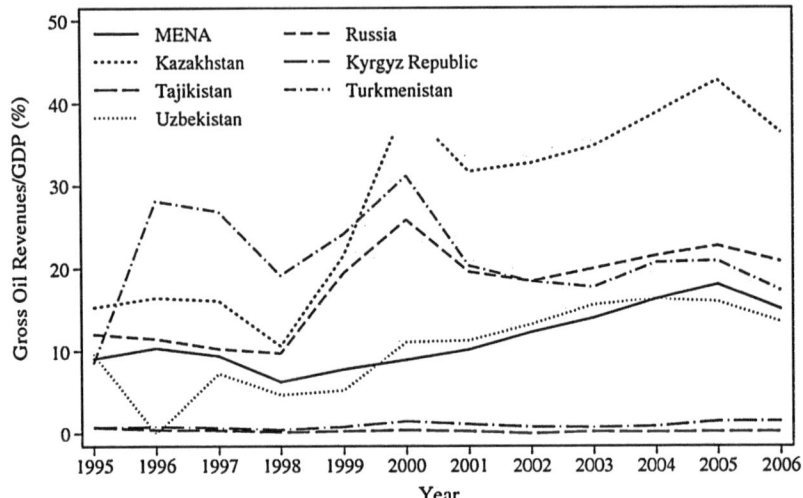

Source: International Energy Agency (2007, IMF (2007).

the colonial rule. Similarly, Beck and Laeven (2006) find that the CEEC were better reformers when compared to their Central Asian counterparts, because they have spent less time under a socialist regime with a non-market based institutional framework. In this case, individuals find it costly to adjust to a new set of rules, such that the old ones stay in place. Looking at Figure 1, these factors seem to explain initial values in the graph for Central Asia, CEEC, and the MENA economies. But they do not appear to explain the changes in the reform process over time. MENA economies show higher values at the beginning of the sample when compared to Central Asian countries possibly due to a different legal system implemented during colonial times. But despite more favorable starting conditions MENA countries showed little change throughout the following years and did not manage to catch up with the EU. Moreover, it does not explain why several Central Asian countries have overtaken MENA economies.

As neither history nor geographical factors can be undone, it is important to understand the transmission channels through which historical factors affect current choices. Social norms, habits, or costs arising from adjusting to a new set of rules are likely reasons why agents prefer a status quo or gradual changes rather than a quick adjustment, even if it would be socially desirable. Moreover, economic and institutional reforms are often not Pareto-optimal, such that certain groups in the population will be confronted with utility losses due to these reforms. Even though reforms may be socially optimal on a global scale, the non-existence of a political Coase-Theorem, such that winners cannot credibly commit themselves to compensate the losers of the reforms, makes reforms less likely (Acemoglu 2003; Hoff and Stiglitz 2008). Continuing with a reform path that is sub-optimal for a strategically important group of the voting population or special interest groups bears the risk for politicians of losing their incumbency (Acemoglu and Robinson 2006).

In a similar fashion, individuals' risk preferences may yield a similar effect. Uncertainty about the outcome of reforms, due to the interdependence of different institutions, may also prevent reform efforts or even worsen existing arrangements unintentionally. Hence, the effect of uncertainty on an individual's utility may provoke risk averse agents to prefer keeping their status quo, rather than voting for reforms (Fernandez and Rodrik 1991). As a result, risk averse policy makers lack the willingness to reform. Moreover, individuals may have preferences about the way institutions actually should be designed. It is obvious that these preferences are shaped by either the former institutional system or severe economic shocks and that historically shaped preferences have an effect on current political outcomes. De Grauwe (2007) e.g. argues that historical events explain the differences between the French and German inflation target prior to the introduction of the European Monetary System (EMS). Severe inflationary crises in German history and painful deflationary episodes in France have shaped preferences among politicians and citizens in favor of more stability-oriented, or lax monetary policy, respectively. Similarly, Alesina and Fuchs-Schuendeln (2007) provide evidence on how persistent preferences about institutions are after a regime change by using the division of Germany as a natural experiment. In their paper, Alesina and Fuchs-Schündeln (2007) find that East Germans, who grew up under the socialistic regime, are more in favor of redistribution and state intervention than West Germans. This effect is particularly strong for older cohorts, which have spent more time under the Communist regime compared to younger cohorts. From their results, they estimate that it takes approximately 30 years until preferences between east and west converge.

The education system is another important factor causing persistence of institutions in Central Asia that may be responsible for persistent institutions. Sound property rights protection and well-functioning factor markets are essential preconditions for developing a knowledge-based economy and economic growth driven by ideas. However, having a well-developed property rights protection alone does not ensure production ideas, as it requires a certain level of education among the work force to develop innovations.

Figure 3 shows third level education enrolment rates for Central Asia and the CEEC. There appears to be no relationship between education and institutions around 1995. Kazakhstan and Uzbekistan lead the figures, followed by the CEEC, Turkmenistan, Tajikistan, and the Kyrgyz Republic in the last spot. But third level education enrolment rates roughly mirror the ranking of institutional quality in Figure 1, when looking at 2006 figures. CEEC have the highest share of population with third level education in 2006, followed by Kazakhstan and the Kyrgyz Republic as the countries with the highest share of third level education among Central Asian countries, whereas Turkmenistan and Uzbekistan have the lowest fraction. Hence, education did not seem to matter much in the early and mid-1990s. But education of the labor force seem to be associated with the reform process in recent years, such that investing in education and human capital increases the preference for property rights protection and more market-based institutional arrangements, in order to ensure that individuals can benefit from their human capital accumulation.

Figure 3: Third Level Education Enrolment Rates 1995–2006

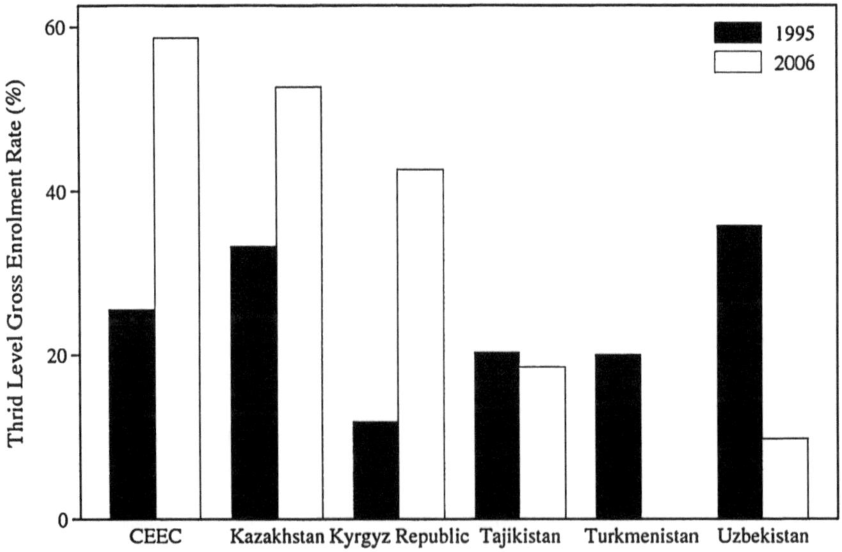

Note: No data available for Turkmenistan after 1997.
Source: World Bank (2008).

A second factor related to education which can be linked to persistence of institutional settings is the quality of education. Although the quality of an education system is hard to measure, a large number of public and private employees in Central Asia, still in key positions after the breakdown of the socialistic regime, were educated under the socialistic system. It is obvious that a curriculum in socialistic book-keeping, banking supervision, and legal treatment of property rights differed remarkably from what a market-based economy demands. In this case, persistence may simply arise from the complexity of setting-up market based institutions and policy makers' lack of knowledge about how to implement and execute certain institutional arrangements. The accumulated knowledge of these employees, therefore, does not assist them in executing their tasks and probably makes redesigning institutions impossible.[15]

3. OPENNESS, COMMITMENT, AND ECONOMIC REFORMS

So far, preferences of individuals and politicians, alongside the deficiencies in the education system, appear to be major obstacles in the reform process in Central Asia, whereas the resource curse does not seem to apply to Central Asian countries. Reforming the education system might help the reform process. But educational reform efforts are subject to the same problem as institutional reforms themselves, namely a lack of willingness to reform. Because educational reforms will only pay off in the long-run, politicians are likely to favor reform projects that will pay off in the short-run. Moreover, changing the domestic demand for better institutions does not ensure that better institutions will be introduced, as the political and economic elite may want to block these reforms.

In this section, I argue that economic openness and factors that are related to it, have played an essential part in reforming institutions in Central Asia since the breakdown of the socialistic systems and are also the reason for why CEEC economies have been even more successful reformers in the past.

Alesina and Fuchs-Schündeln (2007) argue that preferences about institutions in East and West Germany will converge over time as a result of cultural and economic exchange within a highly integrated economy. While there certainly is an effect on preferences due to economic exchange between Central Asian countries and the rest of the world, I focus more on the disciplining effect external factors have on the policy agenda in small open economies and thereby help to overcome reform-unwillingness. A number of papers have underscored the importance of commitment devices in order to cope with time-inconsistency and credibility problems in individual decisions and policy making (Schelling, 1960; Barro and Gordon, 1982 Kydland and Prescott, 1997; Benjamin and

---

15 Vaclav Klaus (1990) summarized this as "When we want to play chess, we must knowhow to play."

Liabson 2003). Commitment devices can work twofold: first, they encourage policy makers not to sacrifice reform efforts that would be beneficial in the long-run for short-term policy objectives. Second, they also signal that already conducted reforms will not be reversed.

Countries with a high degree of openness have a larger share of revenues per GDP from external trade and a larger share of foreign investment in total investment. Therefore, opportunity costs in terms of forgone business opportunities as a result of bad governance provide an incentive to improve institutional arrangements.

Figure 4 shows the degree of trade openness measured as total imports and exports in goods and services as a percentage of GDP in Central Asia and CEEC. Figure 4 shows that CEEC, Kazakhstan, and the Kyrgyz Republic have gradually opened up over time, while trade as a share of GDP has declined in Turkmenistan, Uzbekistan, and especially in Tajikistan. Comparing the changes over time rather than absolute levels in 1995 and 2006, this reflects approximately the ranking in Figure 1.

Figure 4: Trade Openness 1995–2006

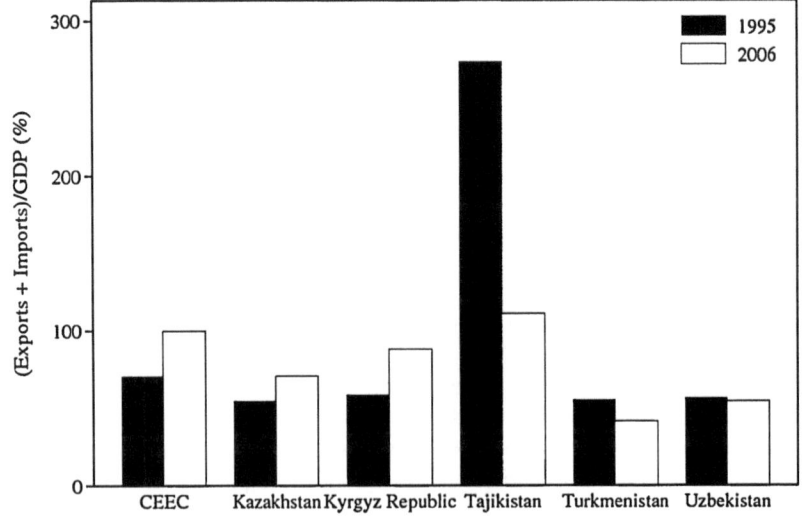

Source: IMF (2008).

A similar argument can be made with regard to financial openness. Lane and Milesi-Ferretti (2006) report that, in the early 1990s, the CEEC have gradually liberalized their capital accounts, which has led to a number of reforms that strengthened financial sector development in these countries. The disciplining effect of financial openness stems from the fact that financial openness not only improves access to international capital markets and foreign investment, but it also comes at the risk of sudden stops and sudden reversals in international capital flows and thus, painful economic crises. Therefore, it provides an incentive to improve the regulatory framework in order to avoid recessions, as well as reducing sovereign risk and borrowing costs abroad.

Figure 5 shows financial openness measures for Central Asian economies and the CEEC as a benchmark from 1995 to 2004. Estimated stocks of total foreign assets plus liabilities as a share of GDP is provided by Lane and Milesi-Ferretti (2007). The data is adjusted for exchange rate and valuation effects in order to provide a clean measure of exposure to international financial markets. Figure 5 shows that the CEECs and Central Asian countries had similar levels of financial integration around 1995, except for Turkmenistan and Uzbekistan. The CEEC show a constant upward trend over the 1990s and early 2000s, similar to Kazakhstan and the Kyrgyz Republic. While all Central Asian countries, except for Uzbekistan, overtook CEEC for a short period in the following years, only Kazakhstan and the Kyrgyz Republic sustained high levels of financial openness similar to the CEEC after the year 2001, which coincides with the changes documented in Figure 1. Kazakhstan and the Kyrgyz Republic already had a higher level of institutional quality when compared to their peers in Central Asia. But they also continued to improve over the following years. In contrast to this, Turkmenistan stopped in particular improving institutions (Figure 1), which went hand in hand with financial disintegration (Figure 5).

Financial openness, per se, presumably does not provide an incentive to improve regulatory frameworks. Giavazzi and Pagano (1988) have argued that fixed exchange rate rates, as opposed to flexible ones, also have a disciplining effect on the policy agenda in small open economies. Fixed exchange rates are often a necessity in transition countries, because of the inability to borrow in domestic currency and to hedge the exchange rate risk due to underdeveloped capital markets, as well as the inability to conduct stability-oriented monetary policy autonomously (Calvo and Reinhart 2002).

On the other hand, fixed exchange rates provide a target for speculative attacks. In the event that agents expect undesirable policies, they are likely to speculate against the country's currency, which may force monetary authorities to abandon the peg. As a result, a rapid devaluation will trigger a default of the

Figure 5: Financial Openness 1995–2006

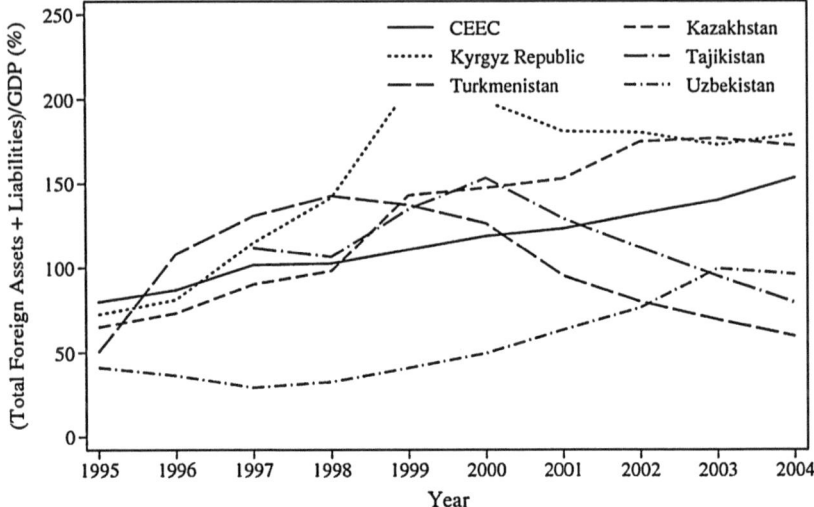

Note: No data available for Tajikistan before 1997.
Source: Lane and Milesi-Ferretti (2007).

government and firms on foreign currency loans which, in the medium and long-run, will have negative consequences on the government's and firms' ability to refinance on international markets. Thus, tying the government's hands by fixing the exchange rate is likely to change priorities on the policy agenda in favor of reforms towards more market-based institutions, such as sound financial regulation and property rights protection, in order avoid speculative attacks on the exchange rate regime.

Figure 6 shows average annual exchange rate volatilities of national currencies from 1995 to 2006 against the US dollar (USD), the euro, and the German Mark (DM) prior to the introduction of the euro. Exchange rate volatilities are calculated as Z-Scores, as proposed by Ghosh et al. (2002). This allows us to summarize the de-facto exchange rate regime regardless of the actual anchor currency. For each country, the volatility measure z_{jt} against the *j-th* foreign currency at time *t* is calculated as $z_{jt} = \min(\sqrt{\mu^2_{\Delta ejt} + \sigma^2_{\Delta ejt}})$, whereby $\mu_{\Delta e}$ and $\sigma_{\Delta e}$ represent the mean and the standard deviation of the currency returns against the *j-th* currency (USD or euro/DM) in year *t* respectively. The minimum volatility is chosen in order to identify the manipulated and policy relevant exchange rate. Annual figures are calculated from monthly data for each year.

Figure 6 shows that the CEEC have constantly kept exchange rate volatility at very low levels while most Central Asian countries show significantly higher volatility between 1995 and the year 2000 but appear to be on a downward trend. Following the year 2000, Kazakhstan, the Kyrgyz Republic, and Tajikistan have converged to CEEC volatility levels. In contrast to this, Turkmenistan and especially Uzbekistan still show significantly higher levels of exchange rate volatility, which peaks in the aftermath of the Russian debt crisis in 2001. One conclusion that can be drawn from Figure 6 is that the countries with the lowest levels of institutional quality according to Figure 1, Turkmenistan and Uzbekistan, were also those who seemed to suffer the most from contagion effects in the aftermath of the Asian and the Russian crisis in 1997 and 2000. Uzbekistan, in particular, had severe problems in bringing exchange rate volatility back down to pre-crisis levels. Other Central Asian economies that were already closer to the CEEC, in terms of institutional quality, did not have to abandon their exchange rate pegs. Moreover, once the reform process seemed to have speeded up, Uzbekistan was able to contain exchange rate volatility and bring it back to pre-crisis levels.

Figure 6: Exchange Rate Volatility 1995–2006

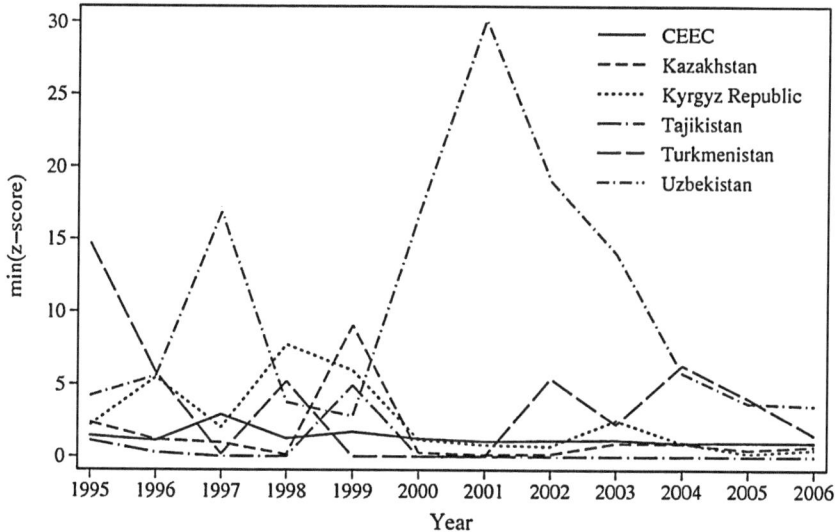

Source: IMF (2008).

Even though other Central Asian economies did not have to abandon their exchange rate regimes during the Russian debt crisis in 2001, which also hit other countries in the region and the CEEC, Figure 1 shows that there were increased reform efforts in Central Asia after 2001. While the incentive of financial openness, in combination with fixed exchange rate regimes, assists in preventing an economic crisis, as shown in the Uzbek case, economic crises themselves can have a positive effect in the long-run, despite painful short-run effects. As already pointed out in the previous section, preferences are often shaped by historical events. Periods of systematic instability can shift preferences in favor of better regulations among the population and policy makers and open a window of opportunity for reforms as they reveal weaknesses in institutional arrangements (Acemoglu and Robinson 2001; Brückner and Ciccone 2011). In the aftermath of the crisis, agents have a higher willingness to reform and prefer better institutions, in order to avoid future costly recessions.

Beside these market factors, international agreements that specifically target economic improvement may provide a form of commitment as well. A significant amount of studies find that IMF and World Bank programs that specifically target political and social change have negative effects on institutions and economic growth (Przeworski and Vreeland 2000; Dreher and Rupprecht 2007). In contrast to this, Di Tomasso *et al.* (2007) find a positive effect of EU or NATO membership on institutional development. The main difference between the two findings can be attributed to the different types of conditionality that apply when it comes to membership and assistance. In the former case, assistance is given in exchange for the promise of political change. In the latter, becoming a member of either NATO or the EU requires political change first and the reward in form of external security and access to European markets comes second.

The CEEC have been the most successful reformers among the former communist economies. A major difference between CEEC and other emerging markets at the EU periphery is that the countries in Eastern Europe have a reasonable prospect of entering the EU at some point in the future. There is evidence that the prospect of an EU membership has become an anchor for domestic economic policy-making in many CEEC (Babetskii *et al.* 2004). A (prospective) EU membership imposes important constraints on national fiscal and monetary policy, as well as on other policy areas such as governance, as compliance with the *Acquis Communautaire* is a key requirement for entering the EU (European Commission (EC) 2007). Thus, the prospect of entering the EU is likely to explain the transition observed in the CEEC. Although weaker, other forms of cooperation between the EU and countries at the periphery may also provide a disciplining effect on institutional reforms. Economies in Central Asia (as well as Russia) are part of the TACIS-programme. The EU has underscored the strategic importance of its neighbors in Central Asia and has already become an important export market for members of the TACIS-programme (European Commission (EC) 2007). Although the TACIS-programme is not a political agreement in the first place and resembles more an IMF-type programmes, it has become the corner stone of economic and political integration between the EU and Central Asia (EC 2007). Giving full access to European markets constitutes a strong economic incentive for Central Asian economies to give in to certain political demands from the EU. Danne (2011) provides empirical evidence that the TACIS has helped the institutional transformation process in Central Asia. While there is no direct incentive, as there is in the case of a (prospective) EU membership, there is some evidence that regional non-trade related agreements with a regional hegemon in the center are an important channel through which institutional arrangements diffuse to neighboring countries (Danne 2011).

4. CONCLUSION

This paper compares the institutional reform processes in Central Asian economies to those of their neighboring countries in CEEC and MENA economies. The paper identifies shortcomings in education and preferences about reforms of individuals and policy makers as some the main obstacles in the reform process. Based on this, I identify external factors that have acted as commitment devices for Central Asian economies over the past 15 years to reform existing institutional arrangements. Stylized evidence is provided that external factors, such as real and financial openness and factors that are related to both, provide an incentive for policy makers to sustain institutional change. Real and financial openness, fixed exchange rates, and non-trade related international agreements have disciplined governments to conduct institutional reforms. This results from the fact that small open economies, by definition, are more exposed to external shocks and international capital and trade flows. A high degree of trade and financial openness may provide an incentive to reform institutional settings as small and open countries have a larger share of revenues per GDP from external trade and a larger share of foreign investment in total investment. Therefore, opportunity costs in terms of forgone business opportunities as a result of bad governance provide incentives to improve the institutional set-up. Similarly, financial openness provides an incentive to remedy deficiencies in the regulatory framework, thereby making the economy less prone to sudden capital outflows that may result in financial turmoil and painful economic recessions. Pegging the exchange rate, in combination with financial openness, has the same effect as undesirable domestic policy measures and may result in speculative attacks against the currency regime. External agreements, such as the TACIS-programme are also likely to act as an incentive for Central Asian countries, as institutional reforms towards more market-oriented institutions are essential for closer ties with the EU. Lastly, there is also some evidence that external shocks, such as an economic crisis may shift preferences of individuals and politicians towards better institutional arrangements.

REFERENCES

Acemoglu, D. (2003), 'Why Not a Political Coase Theorem? Social Conflict, Commitment and Politics', *Journal of Comparative Economics*, 31, 4, pp. 620–652.
Acemoglu, D. and Robinson, J.A. (2001), 'A Theory of Political Transitions', *American Economic Review*, 91, 4, pp. 938–963.
Acemoglu, D. and Robinson, J.A. (2006), 'Economic Backwardness in Political Perspective', *American Political Science Review*, 100, 1, pp. 115–131.
Acemoglu, D., Johnson S. and Robinson, J.A. (2001), 'The Colonial Origins of Comparative Development: An Empirical Investigation', *American Economic Review*, 91, 5, pp. 1369–1401.
Acemoglu, D., Johnson S. and Robinson, J.A. (2005), 'Institutions as the Fundamental Cause of Long-Run Growth', in: Aghion, P. and Durlauf, S.N. (eds.), *Handbook of Economic Growth*, North-Holland: Elsevier, pp. 385–464.
Alesina, A. and Fuchs-Schündeln, N. (2007), 'Good Bye Lenin (or Not?): The Effect of Communism on People's Preferences', *American Economic Review*, 97, 4, pp. 1507–1528.
Babetskii, I., Boone, L. and Maurel, M. (2004), 'Exchange Rate Regimes and Shocks Asymmetry: The Case of the Accession Countries', *Journal of Comparative Economics*, 32, 2, pp. 212–229.
Barro, R. and Gordon; D. (1982), 'Positive Theory of Monetary Policy in a Natural Rate Model', *Journal of Political Economy*, 91, 4, pp. 589–610.
Beck, T. and Laeven, L. (2006), 'Institution Building and Growth in Transition Economies', *Journal of Economic Growth*, 11, 2, pp. 157–186.
Benjamin, D.J. and Laibson, D.I. (2003), 'Good Policies and Bad Governments: Behavioural Political Economy', in: Kopcke, R.W., Sneddon Little, J. and Tootell, G.M.B. (eds.), 'How Humans Behave: Implications for Economics and Economic Policy', *Conference Series 48*, Boston, MA.
Brückner, M. and Ciccone, A. (2011), 'Rain and the Democratic Window of Opportunity', *Econometrica*, 2011, 79, 3, pp. 923–947.
Calvo, G.A. and Reinhart, C.M. (2002), 'Fear of Floating', *Quarterly Journal of Economics*, 117, 2, pp. 379–408.
Congdon Fors, H. and Olsson, O. (2007), 'Endogenous Institutional Change after Independence', *European Economic Review*, 51, 8, pp. 1896–1921.
Danne, C. (2011), 'Exiting the Lawless State: An Empirical Investigation', *Mimeo*, Trinity College Dublin.
De Grauwe, P. (2007) *The Economics of Monetary Union*, 7th ed., Oxford: Oxford University Press.
Di Tomasso, M. L., Raiser M. and Weeks, M. (2007), 'Home Grown or Imported? Initial Conditions, External Anchors and the Determinants of Institutional Reform in Transition Economies', *Economic Journal*, 117, 520, pp. 858–881.

Dreher, A. and Rupprecht, S.M. (2007), 'IMF Programs and Reforms – Inhibition or Encouragement?', *Economics Letters*, 95, 3, pp. 320–326.
European Commission (2007), *Enlargement Strategy and Main Challenges 2007-2008*, Brussels: European Commission.
Fernandez, R. and Rodrik, D. (1991), 'Resistance to Reform: Status Quo Bias in the Presence of Individual-Specific Uncertainty', *American Economic Review*, 81, 5, pp. 1146–1155.
Ghosh, A.R., Gulde, A.-M. and Wolf, H.C. (2002), *Exchange Rate Regimes: Choices and Consequences*, Cambridge, MA: MIT Press.
Giavazzi, F. and Pagano, M. (1988), 'The Advantage of Tying One's Hands', *European Economic Review*, 32, 5, pp. 1055–1075.
Hall, R.E. and Jones, C.I. (1999), 'Why Do some Countries Produce so Much More Output than Others?', *Quarterly Journal of Economics*, 114, 1, pp. 83–116.
Heritage Foundation (2008), 'Index of Freedom', online at: http://www.heritage.org/Index/Download.aspx.
Hoff, K. and Stiglitz, J. (2008), 'Exiting the Lawless State', *Economic Journal*, 2008, 118, 531, pp. 1474–1497.
International Monetary Fund (2005), 'World Economic Outlook - Building Institutions', *World Economic Outlook Paper*, Washington, DC.
International Monetary Fund (2007), 'The Poverty Reduction and Growth Facility (PRGF)', *Factsheet*, Washington, DC.
Klaus, V. (1990), 'A Perspective on Economic Transition in Czechoslovakia and Eastern Europe', in: Fischer, S., de Tray, D. and Shah, S. (eds.), *Proceedings of the World Bank Annual Conference on Development Economics*, Washington, DC: World Bank, pp. 13–18.
Kuran, T. (2004), 'Why the Middle East is Economically Underdeveloped: Historical Mechanisms of Institutional Stagnation', *Journal of Economic Perspectives*, 18, 3, pp. 71–90.
Kydland, F.E. and Prescott, E.C. (1977), 'Rules rather than Discretion: The Inconsistency of Optimal Plans', *Journal of Political Economy*, 85, 3, pp. 473–491.
La Porta, R., Lopez-de-Silanes, F. and Schleifer, A. (2008), 'The Economic Consequences of Legal Origins', *Journal of Economic Literature*, 46, 2, pp. 285–332.
Lane, P.R. and Milesi-Ferretti, G.M. (2006), 'Capital Flows to Central and Eastern Europe', *IMF Working Paper 06/188*, Washington, DC.
Lane, P.R. and Milesi-Ferretti, G.M. (2007), 'The External Wealth of Nations Mark II: Revised and Extended Estimates of Foreign Assets and Liabilities, 1970–2004', *Journal of International Economics*, 73, 2, pp. 223–250.
Peters, B.G. (1996), 'Political Institutions, Old and New', in: R. E. Goodin & H. D. Klingemann (eds.), *A New Handbook of Political Science*, Oxford: Oxford University Press, 1996.
Przeworski, A. and Vreeland, J.R. (2000), 'The Effect of IMF Programs on Economic Growth', *Journal of Development Economics*, 62, 2, pp. 385–421.
Rajan, R.G. and Zingales, L. (2006), 'The Persistence of Underdevelopment: Institutions, Human Capital or Constituencies', *CEPR Discussion Paper No. 5867*, Cambridge, MA.
Ramsay, K. (2006), 'The Price of Oil and Democracy', *Mimeo*, Princeton University.
Ross, M.L. (2001), 'Does Oil Hinder Democracy?', *World Politics*, 53, 3, pp. 325–361.
Sachs, J.D. and Warner, A.M. (2001), 'The Curse of Natural Resources', *European Economic Review*, 45, 4-6, pp. 827–838.
Sala-i-Martin, X. and Subramanian, A. (2003), 'Addressing the Natural Resource Curse: An Illustration from Nigeria', *NBER Working Paper 9804*, Washington, DC.

Schelling, T.C. (1960), *The Strategy of Conflict,* Cambridge, MA: Harvard University Press.
Teorell, J., Charron, N., Samanni, M., Holmberg S. and Rothstein, B. (2009), 'The Quality of Government Dataset', University of Gothenburg: The Quality of Government Institute.

APPENDIX

Table A: Definitions of Country Groups

Central Asia   Kazakhstan, Kyrgyz Republic, Tajikistan, Turkmenistan, Uzbekistan
CEEC   Albania, Bulgaria, Croatia, Czech Republic, Estonia, Hungary, Latvia, Lithuania, Macedonia, Poland, Romania, Slovakia
EU-15   Austria, Belgium, Denmark, Finland, France, Germany, Greece, Ireland, Italy, Luxembourg, Netherlands, Portugal, Spain, Sweden, United Kingdom
MENA   Algeria, Egypt, Israel, Jordan, Lebanon, Libya, Morocco, Syria, Tunisia

PART II: Case Studies

The Effects of Foreign Direct Investment on Wages in Kazakhstan

Saule Sagandykova

1. INTRODUCTION

From 1998 to 2006 the world foreign direct investment stock had increased three times and reached $12.5 trillion (UNCTAD 2009). The growth in capital inflows led to a substantial interest in the impact of FDI on economic growth and welfare in host countries. A topic that has sparked controversy is the impact of FDI on wages, as theoretical studies do not provide an unambiguous prediction of the effect of FDI on local wages in developing countries. The effect is therefore an empirical question (Ge 2006; Brown et al. 2003). The central objective of this paper is to investigate how FDI affected wage levels and wage inequality in Kazakhstan before 2007, i.e. the beginning of the global economic and financial crisis.

Since 1993, Kazakhstan has successfully attracted a large amount of FDI and is now one of the three largest recipients of FDI in the Commonwealth of Independent States. However, during 1996-2006, a significant feature of the Kazakh economy was increasing wage inequality, especially between agrarian and petroleum regions and industries (see Table 1 in the appendix). Petroleum rich Caspian regions such as Atyrau, West Kazakhstan, and Mangistau accumulated most of the FDI, and the average real wage level in these regions was the highest in Kazakhstan. A main objective of the Kazakh government's external policy was to attract FDI in order to achieve balanced economic growth and development. Our study provides evidence of a close linkage between the distribution of FDI and wage inequality among regions and industries, and consequently has policy implications in relation to the promotion of FDI.

The wage equation was constructed with panel data on regional and industry differences. An ordinary least squares estimator (OLS) was used. The results suggest that FDI has a significant and positive impact on real wages in Kazakh regions and industries, and that this impact remains significant after controlling for time, regional and industrial characteristics. It was found that cross-regional and cross-industrial variation of FDI intensity is believed to be one of the important factors contributing to wage inequality.

The remainder of this paper is organized as follows. Section 2 reviews the literature of the linkage between FDI and wages. Section 3 defines the relevant data. Section 4 describes the method, and section 5 empirically examines the impact of FDI on wages by regions and industries. Section 6 concludes.

2. THE LINKAGE BETWEEN CHANGES OF FDI AND WAGES

The direction of the causality linkage between the variance of FDI and the variance of wage is ambiguous. On the one hand, host country wage levels may affect the location decision of FDI. It depends on the size of wage (low or high) and can happen in two ways.

Some Multinational Enterprises (MNEs) search for low labour cost for a given labour productivity. In this case, the evidence in the literature is mixed as some studies find a positive correlation between FDI and wages while others find a negative link (Wheeler and Mody 1992; Billington 1991; Head *et al.* 1999). Other MNEs are *cherry pickers*, and they prefer to acquire the most productive and high-wage plants (Almeida 2003; Harris and Robinson 2002). Thus, positive effects of FDI on wages result from companies' specific preferences.

On the other hand, the linkage between FDI and wages can also go in the opposite direction. First, FDI can influence wages in private and state[16] sectors through the following two channels (Ge 2006) (1) the "wage differential effect" (MNEs pay higher or lower wages in their companies in comparison with domestic companies) and (2) the "wage spillover effect" (MNEs affect the wage levels of local companies via spillover of knowledge and technology). Secondly, MNEs can indirectly influence wages in the state sector through financial relationships with governmental (national and local) budgets. These relationships include payments to budgets (taxes, custom duties) and receiving different kinds of promotion from the budgets (tax and custom exemption, tax holidays, natural grants, and other incentives and subsidies).

2.1 The wage differential effect

The majority of empirical evidence supports the claim that firms with a foreign ownership share (joint venture or foreign companies) pay higher wages than domestic firms, even after the effect of scale, labour quality, industry, regional location and some other characteristics are taken into account. The estimated wage gaps between MNEs and local firms varied from 6 to 260 per cent in developed (the United States, the United Kingdom) and developing (Morocco, Venezuela, Indonesia, Cameroon, Ghana, Kenya, Zambia and Zimbabwe) countries in different geographical locations during the last two decades (Haddad and Harrison 1993; Lipsey, 1994; Aitken *et al.* 1996; Harrison 1996; Feliciano and Lipsey 1999; Griffith and Simpson 2001; Girma *et al.* 2001; Driffield and Girma 2003; Choi 2003; Velde and Morrissey 2003, Lipsey and

---

16 The state (or public) sector in this study includes companies and organizations, which are financed from governmental budget on national and local levels.

Sjoholm 2004; Taylor and Driffield 2005). Velde (2003) also found that foreign establishments in Britain paid higher wages than domestic establishments. However, the differential disappeared when the skill structure within establishments was controlled.

2.2 The wage spillover effect

The empirical evidence for the wage spillover effect reveals mixed results. Girma et al. (2001) studied FDI in the United Kingdom and found no overall wage spillover effect on wage levels, but a negative effect on wage growth. Barry et al. (2005) found that a foreign presence has a negative effect on wages and productivity in domestic exporting firms in Ireland, possibly due to the labour market crowding-out effect. Driffield and Girma (2003) found a positive wage spillover effect, but this effect was confined to the region where FDI takes place. Taylor and Driffield (2005) found that FDI had a significant effect on wage dispersion which can be interpreted as evidence for a technology spillover.

2.3 Impact on state wages

Taxes and customs fees paid by the MNEs can raise opportunities to increase wages in the public sector. However, tax exemptions and financial subsidies, which MNEs also receive, can outweigh the positive effects of FDI and can limit governmental incomes. Figlio (2000) found that while increased manufacturing employment increases industrial wages in a country, the effect was more than seven times larger when the employment growth came from a foreign firm, rather than a domestic one. However, on the budget side, he found that foreign employment leads to a larger decline in per capita revenues and to expenditures at the county level in South Carolina. Hanson (2001) found some evidence that a host country's subsidies to foreign firms might actually lower a country's welfare.

In various studies the effects of FDI on wages are considered on the following interrelated levels: the labour force, the firm, the region and industry, and the country level.

2.3.1 Labor force

Most MNEs seek to attract high quality labour. The workers have to be educated, qualified, skilled, and experienced. The level of qualification strongly influences the wage level. There is supporting evidence, for both industrialized and developing countries, that FDI increases wage inequality between skilled and unskilled workers, as well as for a skill-premium for workers with higher education. Positive direct impact of FDI on the skill-premium has been found in Hungary and in the Czech Republic. These countries received the largest FDI inflows in Central and Eastern Europe in the 1990s and experienced rapid increases in wage inequality between skilled and unskilled workers (Bruno *et al.* 2004).

FDI can decrease or increase wage inequality based on population characteristics such as race, sex, and position. Gelan *et al.* (2007) indicate that FDI is associated with an improved probability of United States black employment in high-wage and mid-level wage jobs versus low-wage jobs. These findings suggest that FDI creates greater job opportunities for individuals from groups traditionally under-represented in high-paying occupations. Vijaya and Kaltani (2007) analyzed nineteen developing and developed countries, and their results indicate that FDI flows had a negative impact on overall wages in the manufacturing sector and that this impact was stronger for female wages. Lipsey and Sjoholm (2004) studied wages in Indonesia and showed a wage premium of 12% for blue-collar workers and 22% for white-collar workers. Foreign-owned firms were found to pay a higher price for labor of a given educational level than domestically owned ones.

2.3.2 Firm

Size of establishments, internal investments and wage policies of MNEs are usually found to be important determinants of average pay. Large firms pay more for observationally equivalent workers than small firms (Polachek and Siebert 1993).

The investment policy of MNEs could result in a broad spectrum of effects related to whether FDI is carried out through a greenfield investment or through mergers and acquisitions of local firms (Choi 2003). In the first case, MNEs could increase the demand for skilled or/and less skilled workers, and accordingly raise wages. In the second case their impact on local wages is not clear. A positive wage effect of acquisitions has been found in the United States (Brown and Medoff 1988; Nguyen *et al.* 1995). A possible explanation for a positive effect on wages from takeovers could be caused by strategies of new owners to keep personnel and their firm-specific knowledge in the company and to perform organizational changes (Shleifer and Summers 1988; Bertrand and Mullainathan 2003; Heyman *et al.* 2007).

Hunya and Geishecker (2005) suggested that the nature of FDI in manufacturing will remain to be low-wage seeking, vertical, and export-oriented. Galgoczi (2003) reported that, according to an analysis of the Hungarian Metal Unions,

MNEs matched their wage and welfare policies solely to the local conditions. They do not necessarily transfer their employment practices and industrial relations to the host countries.

2.3.3 Region and Industry

Some studies have examined the effect of FDI on wages based on regional and industrial characteristics. On the regional level, FDI location and impact can depend on region specialization, urbanization, and openness (open economic zones). For example, Ge (2006), using a panel dataset of Chinese cities, examined the impact of inward FDI on urban real wages. His results showed that the existence of FDI has a significant and positive effect on urban real wages and that this impact remains significant after controlling for other city characteristics.

On the industrial level the effect can be very different for capital intensive versus non-capital intensive industries, and for manufacturing versus non-manufacturing industries, etc. Positive effects of FDI on wages in countries such as the Czech Republic, Hungary, Poland, Slovakia, Slovenia have been observed for the capital and skill intensive sectors only (Onaran and Stockhammer 2008) between 2000 and 2004. Choi (2003) did not find any positive association between inward FDI activities and industry wage premiums within United States manufacturing industries in a panel data analysis. Feliciano and Lipsey (1999) found few signs of the wage spillover effect in American manufacturing industries, but a large and significant effect in non-manufacturing industries.

2.3.4 Country

Characteristics of host and source country influence the FDI impact. Figini and Görg (2006) found that the effect of FDI differs according to the level of development. They describe two different patterns, one for the Organisation for Economic Co-operation and Development (OECD) (developed) and one for non-OECD (developing) countries. Moreover, their results suggest the presence of a non-linear effect in developing countries; wage inequality increased with the FDI stock but this effect diminished with further increases in FDI. For developed countries, wage inequality decreased with FDI inward stock and there was no robust evidence to show that this effect was also non-linear.

Gopinath and Chen (2003) found for a sample of eleven developing countries, that FDI flows widen the skilled-unskilled wage gap (measured as the share of unskilled labour in gross national product) for a subset of developing countries, although they appear to lead to cross-country convergence of wages. Aitken *et al.* (1996) found a lack of wage spillover, but significant wage differentials between foreign and domestic firms in Mexico and Venezuela. In the United States, there is a small wage differential, but some evidence supporting the wage spillover effect. Tsai (1995) studied the link between FDI and inequality using a sample of thirty three developing countries and showed that FDI increased inequality only in some Asian countries.

Girma *et al.* (2001) used a firm level panel data set and found that foreign firms pay five per cent higher wages after allowing for industry, size and productivity effects; if the source country was the United States, firms had the largest differential, and if the source country was Japan, they had the smallest.

3. THE DATA

The impact on wage and wage inequality from FDI in Kazakhstan was estimated using two balanced panel data sets for the period from 1998 to 2006. The first set contained fourteen administrative regions, and the second set contained eleven industries. Information about regions and industry characteristics were obtained from the Kazakh National Agency of Statistics. The industrial data could not be disaggregated to a regional level. Such data was not available, and therefore two separate analyses were performed.

Industry data included nominal wage and number of employees by forms of ownership (state, private, foreign); number of employees, output and services of companies with foreign share of ownership (CFSs); share of regions in industrial production; rate of unemployment; investment in fixed capital by form of ownership. Data of FDI inflows by industries and consumer price index (CPI) was compiled from the National Bank of Kazakhstan. Summary statistics for regional and industrial wages and other factors are reported in Table 2 in the appendix.

The data showed that both FDI and real wages increased during the study period (1998–2006) (Figure 1 in the appendix). To measure FDI and wage inequality Gini coefficients were used. The Gini coefficient is a measure of statistical dispersion with values between 0 and 1 (Corrado 1921). A low Gini coefficient indicates a more equal wage and FDI distribution, while a high Gini coefficient indicates a more unequal distribution. The growth of FDI and wage (Figure 1) do not have any apparent co-existence with FDI and wage Gini coefficients (Figures 2 and 3). While FDI and wages increased (Figure 1) Gini coefficient levels remained relatively stable in regions (Figure 2) and demonstrated the joint decrease in industries occurring in 2006 (Figure 3).

Figure 2 in the appendix shows the changes in Gini coefficients of real wages and FDI from 1998 to 2006 for all regions of Kazakhstan. The plot suggests the co-existence of FDI and wage inequalities. Inequalities rose from 1999 to 2000 when the main privatization process with foreign participations came to an end. At the same period large extractive Kazakh companies had been giving their trust to foreign-company management obliged to invest. Since 2000 the level of wage inequality decreased slowly corresponding to the general process of decreasing FDI inequality. Figure 3 indicates that changes of Gini coefficients of real wages and FDI are associated and they have similar directions of change on an industrial level.

Other factors which can affect real wage levels are national investment (in fixed capital), unemployment rate, productivity, the share of regions in industrial production, and the ratio of employees of CFSs to the total number of employees.

4. ESTIMATION METHODOLOGY

Two panel data regression models with time and entity (region or industry) fixed effects were constructed to estimate the impact of FDI on regional and industrial wage levels and for controlling for omitted variables in panel data when they vary across entities and time. This is done by including both n-1 entity binary variables M and T-1 time binary variables N in the regressions, along with the intercept.

The combined time and entity fixed effects regression models can be written as for region:

(1) $RRW_{rt} = b_0 + b_1 RFDI_{rt} + \sum_{r=2}^{n} a_r M_r + \sum_{t=2}^{T} e_t N_t + (cU_{rt} + dSI_{rt}) + \varepsilon_{rt}$

for industry:

(2) $IRW_{it} = b_0 + b_1 IFDI_{it} + \sum_{i=2}^{n} a_i M_i + \sum_{t=2}^{T} e_t N_t + (cP_{it} + dRE_{it}) + \varepsilon_{it}$,

where r is a region (from 1 to 14); i is an industry (from 1 to 11); t is a year (from 1998 to 2006).

RRW is the average real wage per month in the regions; *IRW* is the average real wage per month in an industry; *RFDI* is the *FDI* inflows by regions; *IFDI* is the *FDI* inflows by industries; *U* is the rate of unemployment by regions; *SI* is the share of regions in industrial production; *P* is the productivity in companies with a foreign share ownership; *RE* is the ratio of employees in CFSs to total number of employees; b_0 is a constant; b_1, $a_2,...a_n$, $e_2...e_T$ are unknown coefficients to be estimated; ε_{rt} is an error term, $M2_r$ and $N2_t$ are binary variables ($M2_r=1$ if $r=2$ and $M2_r=0$ in all other cases; $N2_t=1$ if $t=2$ and $N2_t=0$ in all other cases).

In this study *RRW* and *IRW* are dependent variables, and the objective was to examine the effect of FDI on these variables, and more precisely, if their correlations lead to an increase in welfare and differences among regions and among industries. FDI inflows were chosen as regressors because their variances are highly associated with changes in wage levels in regions and industries. Also two other regressors were used as parameters to detect impacts of FDI on RW: first, total investment in fixed capital (*RTI* – for regions, and *ITI* – for industries), and, second, ratio of FDI to total investment in fixed capital (*RF/T* – for regions, and *IF/T* – for industries). A high level of correlation between FDI and total investment in fixed capital: 0.91 for regions and 0.84 for industries, shown in Table 3, made it necessary to investigate the impact of these factors separately.

Four additional control variables were chosen based on existing studies and available data. The first variable was *U*, which can influence the wage level in regions. An increase in unemployment is associated with reductions of wages. The second variable *SE* was chosen to capture the effect of industry structure (prevalent in petroleum, finance, agriculture, manufacture, service or other sectors).

The next variable, *P*, which is a measure of productivity in CFSs was used in the cross-industry regression. If productivity of CFSs is higher than productivity in domestic companies, it can lead to higher wages in joint ventures and foreign-owned companies and can affect average wage in industries. Finally, an increase in the share of employees working in a company with a foreign ownership share (*SE*) has a positive effect on wages if wages in such companies are higher than in national companies, and should subsequently have a positive impact on host country wages in general.

The panel data set lets to control variables that differ from one entity (region or industry) to the next, such as prevailing cultural attitudes, geographical location, natural resources, development of industrial infrastructure, business environment, but do not change over time. It also allows controlling variables that vary through time, i.e. new investment laws and regulations, changes in economic and investment policies, but do not vary across entities.

5. EMPIRICAL RESULTS

Tables 4 and 5 in the appendix summarize results for the OLS regression of FDI inflows on RW. Columns (1), (3), and (5) report the regression solution with entity and time fixed effects; columns (2), (4), and (6) include additional regressors along with the fixed effects. The first two columns ((1) and (2)) represent results of impact on real wages (RRW and IRW) of FDI (RFDI and IFDI), the next two columns ((3) and (4)) show the regression results between real wages and total fixed capital investment (RTI and ITI), and the last two columns ((5) and (6)) show the impact of FDI to total investment in fixed capital on average real wage (RF/T and IF/T).

The results show that all $\bar{R}^2$ are high (Tables 4 and 5). They vary from 0.940 to 0.967 in the regression analysis for Kazakh regions and from 0.888 to 0.945 for Kazakh industries. All region and industry effects are statistically significant at the 1% significance level. SI and U in the regression analysis of the effect of FDI on RW in regions are jointly significant at the 1-5% level. RE and P in the regression analysis of the effect of FDI on RW in industries are jointly significant at the 1% level.

The first column of Table 4 reports that the coefficient on the RFDI is positive (0.04) and has an error of 0.004 meaning that the estimate is significantly different from zero at the 1% level. Accordingly, increasing RFDI increased RRW. Two additional potential determinants (SI and U) along with the region and time fixed effects are added in column (2). Both are positive, and SI is significant. Including the additional variables did not change the estimated coefficient representing FDI impact relative to the regression in column (1). The estimated coefficient (0.04) continues to be positive and statistically significant at the 1% significance level.

The next two columns of Table 4 represent similar results for the regression between RTI and RRW. All estimated coefficients are positive and significant, with the with the exception of U. The impact of U is not significant. These results suggest that the increasing RTI raised wages regionally. Thus, RFDI and RTI both have a positive and significant impact on RRW.

In order to clarify the difference between the impact on wage of foreign and total investment the regressors RFDI and RTI were substituted with a ratio between them (RF/T). The results of this regression (two last columns of Table 4) indicate that coefficients representing RF/T are positive in both cases and they are statistically significant at 1% level. It allows concluding that increasing RF/T increased RRW, and the RFDI impact is more effective than RTI impact.

Table 5 demonstrates significant results on an industrial level. Coefficients in column (1) are positive and statistically significant at the 1% level. They remain positive after adding additional regressors (column 2), such as Re and P. The estimate of IFDI was reduced in comparison with column (1). Both additional coefficients are positive and statistically significant.

The next two columns of Table 5 report the impact of ITI on the IRW level. The results by industry were similar to the results based on the regional level. They show that the impact of ITI on IRW was similar to the impact of RFDI on RRW. Coefficients of the estimate were positive and significant at the 1% level. The coefficients for additional regressors are also positive and significant. Thus, increasing both IFDI and ITI increases IRW in industries.

The results from the analysis between IF/T and IRW (the last two columns of Table 5) are not consistent with the previous results. The estimated coefficient (column (5)) is positive, but not significant. However, it remains positive and becomes significant at the 10% significance level when two additional regressors (RE and P) are included (column (6)). Coefficients related to RE and P are also positive and significant at the 5% and 1% significance level, and the F-statistic is 32.8, when time effects equals zero. Thus, according to these estimate, increasing P and RE along with increasing IF/T were associated with increasing IRW in Kazakhstan during the 1996-2006 period.

Average nominal wages recorded as three forms of ownership (state, private, and foreign) indirectly supports the regression results, and confirms that wages in the foreign sector are higher than wages in state and private sectors. Data on wages in the private ownership sector were not separated into on national and joint venture ownership, and thus provided a limited opportunity to investigate such a difference. However, the difference between foreign and state wages in regions and industries can be an adequate indicator for the current study since foreign and state ownership are associated directly with foreign and state investment.

Figure 4 shows the changes of wage levels in industries. The difference between the mean of the average nominal wage in foreign and state companies increased by 6.1 times. The standard deviation of the average nominal wage was largest in the foreign sector and much smaller in the private and state sectors. This means that foreign ownership allows some groups of employees to receive higher wages than others. For example, the highest average nominal wages in companies with foreign ownership exceeded average wages in companies with state ownership twofold.

Figure 5 illustrates the mean and standard deviation of the avarege nominal wage levels and their growth by forms of ownership in regions. The mean and standard deviation of wages are higher in the foreign sector than in the state and private sectors. During the study period, the difference between the means of foreign and state wages increased by 2.8 times, and the difference between the means of foreign and private wages increased by 1.7 times.

6. CONCLUSION

This study uses regional and industrial data to directly test the impact of FDI on average real wage levels in Kazakhstan. The impact that FDI had on wages was compared with the impact of total investment at the regional and industrial levels. The results suggest a positive and significant link between FDI and average real wages. This positive effect remains significant at the 1% significance level after controlling time and regional effects along with inclusion of additional variables such as U and SI (for regions study); RE and P (for industries study). Similar results were obtained for total investment impact on real wages, however the impact of FDI was larger. The estimate of the RF/T coefficient was positive and significant at the 5% level for inter-regional level and IF/T is positive and significant at the 10% level for inter-industrial level when fixed effects were included. Our results suggest that FDI tend to raise regional and industrial real wage levels. These findings are consistent with those reported in previous studies that link wages and FDI.

The study shows that the average and standard deviation of nominal wages for companies with foreign ownership was higher than for companies with state and private ownership. The increasing number of employees working in the foreign sector raises the overall average nominal wage in the country.

These results are relevant for the Kazakh FDI policy. The present economic policy aims for sustainable economic growth, and it is expected that FDI will provide beneficial effects. This cross-regional and cross-industrial analysis indicates that FDI inflows have positive and significant effects on real wages in Kazakhstan, and that this effect is higher than the effect of other types of investments.

A large share of foreign investment in the total investment stock in regions or in industries for the most part was closely correlated with high wages and vice versa. That is why the concentration of FDI and its impact on wages in extractive industries and in the western petroleum regions (Atyrau, Mangistau, West Kazakhstan), can be one of several competing explanations of rapidly rising inter-regional and inter-industrial wage inequality since 1998.

To avoid a skewed distribution of FDI across regions and industries and to achieve the strategic goal of a sustainable development, the Kazakh government should examine both the investment climate by regions and the industrial priorities of the national investment policy. Appropriate economic, political and other instruments should be developed to determine the best locations of FDI within Kazakhstan.

REFERENCES

Aitken, B., A. Harrison, and R. Lipsey, R. (1996), 'Wages and foreign ownership: a comparative study of Mexico, Venezuela, and the United States,' *Journal of International Economics*, 40, 345-371.

Almeida, R. (2003), *The effects of foreign owned firms on the labour market*. Institute for the Study of Labor. Available from: ftp://repec.iza.org [Accessed 5 August 2008].

Barry, F., H. Görg, H., and E. Strobl (2005), 'Foreign direct investment and wages in domestic firms in Ireland: Productivity spillovers versus labour-market crowding out,' *International Journal of the Economics of Business*, 12 (1), 67-84.

Bertrand, M. and S. Mullainathan (2003), *Are Emily and Greg more employable than Lakisha and Jamal? A field experiment on labor market discrimination*. NBER WP #9873. Available from: http://ideas.repec.org [accessed 5 August 2008].

Billington, N. (1991), 'The location of foreign direct investment: an empirical analysis,' *Applied Economics*, 31, 65-76.

Brown, C. and J. Medoff (1988), 'The impact of firm acquisition on labor, corporate takeovers: Causes and consequences,' in: A. Auerbach, ed. *Corporate Takeovers: Causes and Consequences*. Chicago, IL: University of Chicago Press, 9-25.

Brown, D., A. Deardorff, and R. Stern (2003), *The effects of multinational production on wages and working conditions in developing countries*. The National Bureau of Economic Research. Available from: http://papers.nber.org [accessed 5 August 2008].

Bruno, G., R. Crinò, and A. Falzoni (2004), *Foreign direct investment, wage inequality, and skilled labor demand in EU accession countries*. Centre for Research on Innovation and Internationalization. Available from: www.cespri.unibocconi.it [accessed 5 August 2008].

Choi, M. (2003), *Inward FDI and inter-industry wage differentials in U.S. manufacturing industries*. University of Connecticut, Department of Economics. Available from: http://ideas.repec.org [accessed 5 August 2008].

Corrado, G. (1921), 'Measurement of inequality and incomes,' *The Economic Journal*, 31, 124-126.

Driffield, N. and S. Girma (2003), 'Regional foreign direct investment and wage spillovers: plant level evidence from the UK electronics industry,' *Oxford Bulletin of Economics and Statistics*, 65, 453-474.

Feliciano, Z. and R. Lipsey (1999), *Foreign ownership and wages in the United States, 1987-1992*. NBER WP #6923. Available from: www.nber.org [accessed 5 August 2008].

Figini, P. and H. Görg (2006), *Does foreign direct investment affect wage inequality? An empirical investigation*. Institute for the Study of Labor. Available from: http://ftp.iza.org [accessed 5 August 2008].

Figlio, D. (2000), 'The effect of FDI on local communities,' *Journal of Urban economics*, 48, 338-363.

Galgoczi, B. (2003), 'The impact of multinational enterprises on the corporate culture and on industrial relations in Hungary,' *South-East Europe Review for Labour and Social Affairs*, 6 (1/2), Jul., 27-44.

Ge, Y., (2006), 'The effect of foreign direct investment on the urban wage in China: An empirical examination,' *Urban Studies*, 43 (9), 1439-1450.

Gelan, A., K. Husbands-Fealing, and J. Peoples (2007), 'Inward foreign direct investment and racial employment patterns in U.S. manufacturing,' *American Economic Review Papers and Proceedings*, 97 (2), 378-382.

Girma, S., D. Greenaway, and K. Wakelin (2001), 'Who benefits from foreign direct investment in the UK?' *Scottish Journal of Political Economics*, 48 (2), 119-133.

Gopinath, M. and W. Chen (2003), 'Foreign direct investment and wages: A cross-country analysis,' *Journal of International Trade and Economic Development*, 12, 285-309.

Griffith, R. and H. Simpson (2001), 'Characteristics of foreign owned firms in British manufacturing,' WP #W9573. *Social Science Research Network*. Available from: http://papers.ssrn.com [accessed 5 August 2008].

Haddad, M. and A. Harrison (1993), 'Are there positive spillovers from direct foreign investment? Evidence from panel data for Morocco,' *Journal of Development Economics*, 42, 51-74.

Hanson G. (2001), *Should countries promote foreign direct investment?* Discussion Paper No. 9. UNCTAD. Available from: www.unctad.org [accessed 5 August 2008].

Harris, R. and C. Robinson (2002), 'The impact of foreign acquisitions on total factor productivity: plant level evidence from UK manufacturing, 1987-1992,' *Review of Economics and Statistics*, 84(3), 562-568.

Harrison, A. (1996), 'Determinants of and effects of direct foreign investment in Cote d'Ivoire, Morocco, and Venezuela,' in: M. Robert and J. Tybout, eds., *Industrial Evolution in Developing Countries*. New York: Oxford University Press, 163-186.

Head, K., J. Ries, and Swenson, D. (1999), 'Attracting foreign manufacturing: investment promotion and agglomeration,' *Regional Science and Urban Economics*, 29, 197-218.

Heyman, F., F. Sjöholm, and P. Tingvall (2007), 'Is there really a foreign ownership wage premium? Evidence from matched employer-employee data,' *Journal of International Economics*, 73 (2), 355-376.

Hunya, G. and I. Geishecker (2005), *Employment effects of foreign direct investment in Central and Eastern Europe*. WIIW research reports No. 321. The Vienna Institute for International Economic Studies. Available from: http://wiiw66.wsr.ac.at [accessed 5 August 2008].

Lipsey, R. and F. Sjöholm (2004), 'Foreign direct investment, education and wages in Indonesian manufacturing,' *Journal of Development Economics*, 73, 415-422.

Lipsey, R. (1994), 'Foreign-owned firms and U.S. wages,' NBER WP #4691. Available from: www.nber.org [accessed 5 August 2008].

Nguyen, S., R. McGuckin, and A. Reznek (1995), *The impact of ownership change on employment, wages and labour productivity in US manufacturing 1977-1987*. WP #95/8. Center for Economic Studies, US Bureau of Census. Available from: http://ideas.repec.org [accessed 5 August 2008].

Onaran, Ö. and E. Stockhammer (2008), *The effect of FDI and foreign trade on wages in the Central and Eastern European countries in the post-transition era: A sectoral analysis for the manufacturing industry*. Available from: http://ideas.repec.org [accessed 5 August 2008].

Polachek, S. and W.S. Siebert (1993), *The economics of earnings*. Cambridge: Cambridge University Press.

Shleifer, A. and L Summers (1988), 'Breach of trust in hostile takeovers,' in: A.J. Auerbach, ed., *Corporate takeovers: Causes and consequences*. Chicago, IL: University of Chicago Press.

Taylor, K. and N. Driffield (2005), 'Wage inequality and the role of multinationals: Evidence from UK panel data,' *Labour Economics*, 12 (2), 223-249.

The Base of electronic publications (2009), The Agency of Statistics of the Republic of Kazakhstan. Available from: http://www.stat.kz [accessed 5 November 2008].

Tsai, P. (1995), 'Foreign direct investment and income inequality: Further evidence,' *World Development*, 23 (3), 469-483.

UNCTAD (2009), *Foreign Direct Investment database*. Available from: http://stats.unctad.org/fdi/ [accessed 5 May 2009].

Velde, D.W. te, and O. Morrissey (2003), *Spatial inequality for manufacturing wages in five African countries*. Discussion Paper 2003/66. World Institute for Development Economics Research of the United Nations University. Available from: www.wider.unu.edu [accessed 5 August 2008].

Velde, D.W. te (2003), 'Foreign ownership, microelectronic technology and skills: Evidence for British establishments,' *National Institute Economic Review*, 185 (1), 93-106.

Vijaya, R. and L. Kaltani (2007), 'Foreign direct investment and wages: A bargaining power approach,' *Journal of World-Systems Research*, XIII (1), 83-95.

Wheeler, D. and A. Mody (1992), 'International investment location decisions: the case of US firms,' *Journal of International Economics*, 33, 57-76.

Appendix

The Data Resources on the Internet:

Information for calculation was obtained from
- the Kazakh National Agency of Statistics (http://www.stat.kz) and
- the National Bank of Kazakhstan (www.nationalbank.kz).

Table 1: Foreign investment stock in fixed capital (1998–2006), and real wage (2006) by regions and industries

Region	Foreign investment stock in fixed capital, % of total	Share of foreign investment stock in total investment stock in fixed capital, in %	Real wage, thousands of KZT
Akmola	0.2	3.4	25.5
Aktobe	3.4	10.8	38.2
Almaty	1.7	11.7	27.5
Atyrau	46.6	41.8	68.2
East Kazakhstan	0.2	1.5	30.7
Zhambyl	0.1	3.9	25.0
West Kazakhstan	27.8	74.3	37.1
Karaganda	6.8	27.7	32.1
Kostanai	0.3	4.0	26.9
Kyzylorda	2.9	24.9	33.4
Mangistau	6.4	23.4	66.8
Pavlodar	2.2	16.2	34.3
North Kazakhstan	0.2	5.8	25.0
South Kazakhstan	0.5	4.5	25.3
Total	100		
Industry			
Agriculture, hunting, forestry and fishery	0.1	3.8	17.3
Extractive industry	46.6	59.0	70.0
Processing industry	9.6	32.3	40.2
Electric power, water and gas distributing	1.0	20.2	34.1
Construction	1.9	20.0	51.3
Commerce, car and household appliance repair	4.1	13.3	37.1
Hotels, restaurants	0.2	23.7	44.9
Transport, communication	2.1	24.0	54.3
Finance	1.8	11.2	89.9
Real estate business, leasing and services provided to companies	31.8	62.0	55.8
Education, health care, social services	0.6	12.0	39.7
Total	100		

Table 2: Summary statistics[17]

Region	Variable	Mean	Std. Dev.	Min	Max
RNW	126	21.7	13.5	6.2	74.7
RNWS	126	14.6	5.8	6.7	30.1
RNWP	126	25.4	17.0	5.0	88.3
RNWF	117	46.2	36.6	10.5	180.6
RRW	126	20.2	12.6	6.1	68.3
RFDI	117	22.5	62.5	0.0	442.3
RTI	126	68.4	107.3	1.1	727.6
RF/T	117	0.2	0.2	0.0	1.0
U	126	10.5	2.7	6.9	16.1
SI	126	6.6	5.3	0.7	22.2
Industry					
INW	99	28.7	18.2	3.9	97.5
INWS	99	26.2	23.5	5.2	129.3
INWP	99	28.3	17.3	3.8	96.4
INWF	97	62.5	57.3	6.3	256.2
IRW	99	26.6	16.9	3.6	89.9
IFDI	99	61.8	134.6	-0.3	713.9
ITI	99	104.5	152.7	0.9	660.9
IF/T	88	0.3	0.2	0.0	1.0
RE	88	9.9	11.5	0.0	37.5
P	83	2.9	7.0	0.0	40.3

Table 3: Correlation matrix

Region						Industry				
	RFDI	RTI	RF/T	U	SI		IFDI	ITI	IF/T	P	RE
RRW						IRW					
RFDI	1.00					IFDI	1.00				
RTI	0.91	1.00				ITI	0.84	1.00			
RF/T	0.62	0.44	1.00			IF/T	0.59	0.53	1.00		
U	-0.12	-0.23	0.03	1.00		P	0.56	0.68	0.37	1.00	
SI	0.43	0.58	0.34	0.03	1.00	RE	0.50	0.53	0.49	0.64	1.00

---

17 See meaning of abbreviations in Table 6.

Table 4: Regression analysis of the effect of FDI on average real wages in regions

Dependent variable: average real wage per month (KZT per working person)								
Regressor	(1)	(2)		(3)	(4)		(5)	(6)
RFDI	0.04*	0.04*	RTI	0.04*	0.04*	RF/T	9489.2*	10837.1*
	(0.004)	(0.005)		(0.005)	(0.003)		(3209.4)	(4380.3)
SI		861.8*			748.4*			775.1*
		(254.0)			(156.2)			(261.5)
U		348.6			436.2**			74.6
		(250.0)			(178.3)			(356.5)
F-statistics and p-values testing exclusion of groups of variables:								
Time effects = 0	70.16 (<0.001)	47.53 (<0.001)		60.72 (<0.001)	57.51 (<0.001)		59.01 (<0.001)	32.74 (<0.001)
Region=0	53.73 (<0.001)	26.44 (<0.001)		32.63 (<0.001)	35.16 (<0.001)		39.46 (<0.001)	21.26 (<0.001)
SI, U = 0		6.25 (<0.001)			17.80 (<0.001)			4.40 (0.015)
$\bar{R}^2$	0.940	0.957		0.948	0.967		0.926	0.938

Note: The individual coefficients are statistically significant at the *1%, **5% or ***10% significance level, respectively.

Table 5: Regression analysis of the effect of FDI on average real wages in industries

Dependent variable: Average Real Wage a month (KZT per working person)								
Regressor	(1)	(2)		(3)	(4)		(5)	(6)
IFDI	1.91*	1.71*	ITI	0.02*	0.02*	IF/T	5528.6	7312.0***
	(0.58)	(0.53)		(0.007)	(0.005)		(4310.7)	(3411.7)
		3.19						
RE		0.77**			0.71**			0.89**
		(0.31)			(0.31)			(0.31)
P		0.46*			0.32*			0.50*
		(0.09)			(0.08)			(0.12)
F-statistics and p-values testing exclusion of groups of variables:								
Time effects = 0	31.14 (<0.001)	28.40 (<0.001)		15.84 (<0.001)	19.93 (<0.001)		41.36 (<0.001)	32.83 (<0.001)
Industry= 0	24.68 (<0.001)	51.03 (<0.001)		21.31 (<0.001)	9.16 (<0.001)		30.39 (<0.001)	12.57 (<0.001)
RE, P = 0		14.07 (<0.001)			9.16 (<0.001)			12.57 (<0.001)
$\bar{R}^2$	0.888	0.947		0.894	0.953		0.899	0.945

Note: The individual coefficients are statistically significant at the *1%, **5% or ***10% significance level, respectively.

The Effects of Foreign Direct Investment on Wages 55

Table 1: Abbreviations

MNEs	Multinational enterprises
KZT	Kazakhstan tenge
FDI	Foreign direct investment
RW	Average real wage a month
CFSs	Companies with foreign share of ownership
CPI	Consumer price index
OECD	Organisation for Economic Co-operation and Development

Region

RNW	Average nominal wage/month in regions, thousands of KZT
RNWS	Average nominal wage/month in regions, state ownership, thousands of KZT
RNWP	Average nominal wage/month in regions, private ownership, thousands of KZT
RNWF	Average nominal wage/month in regions, foreign ownership, thousands of KZT
RRW	Average real wage/month in regions, thousands of KZT
RFDI	Foreign investment in fixed capital in regions, billions of KZT
RTI	Total investment in fixed capital in regions, in fixed prices, billions of KZT
RF/T	Ratio of foreign investment to total investment in fixed capital in regions
SI	Share of regions in industrial production, per cents
U	Rate of unemployment in regions, per cents

Industry

INW	Average nominal wage/month in industries, thousands of KZT
INWS	Average nominal wage/month in industries, state ownership, thousands of KZT
INWP	Average nominal wage/month in industries, private ownership, thousands of KZT
INWF	Average nominal wage/month in industries, foreign ownership, thousands of KZT
IRW	Average real wage/month in industries, thousands of KZT
IFDI	Foreign direct investment inflows in industries, billions of KZT
ITI	Total investment in fixed capital in industries, in fixed prices, billions of KZT
IF/T	Ratio of foreign direct investment to total investment in fixed capital in industries
RE	Ratio of number of employees in companies with foreign share of ownership to total number of employees in industries, per cents
P	Productivity in enterprises with foreign share in industries, thousands KZT per worker

Figures

Figure 1: FDI and wage growth

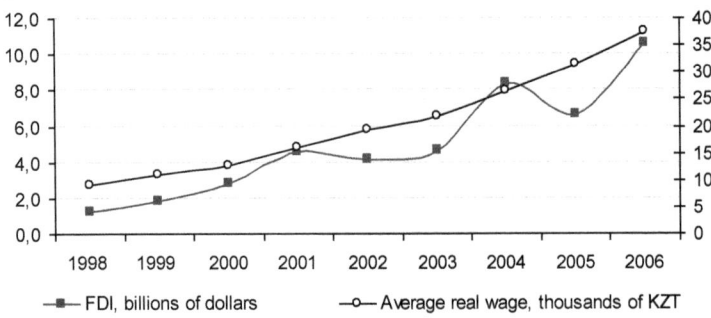

Figure 2: Gini coefficients of wage and FDI in all regions

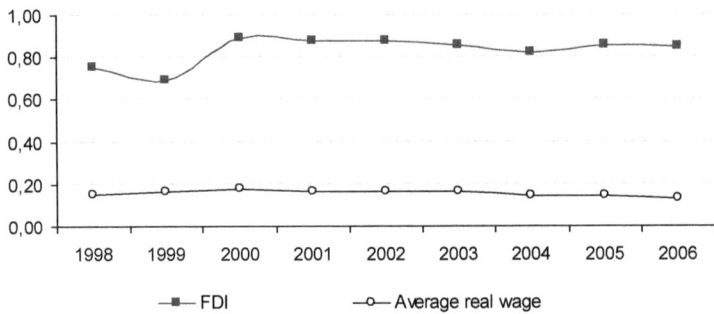

Figure 3: Gini coefficients of wage and FDI in all industries

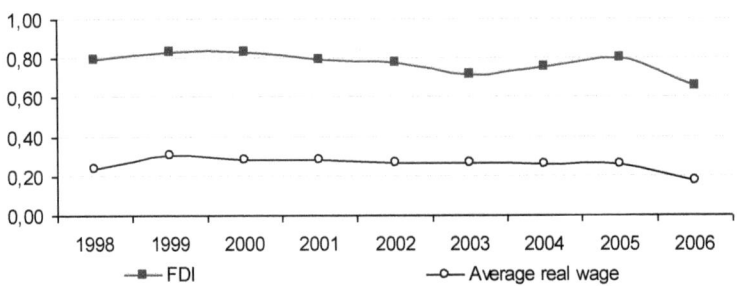

Figure 4: Mean and standard deviation of average nominal wages by forms of property in industries

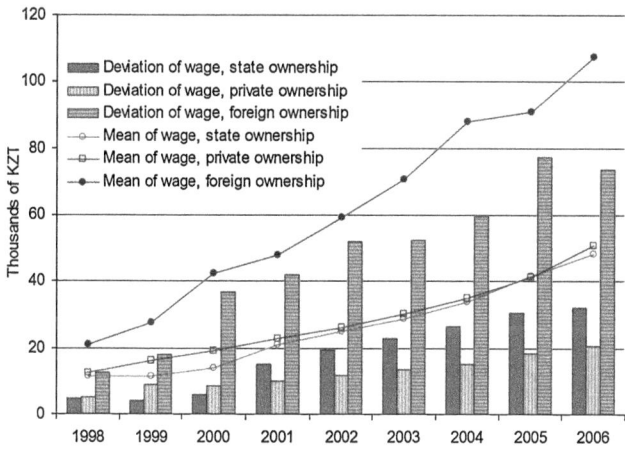

Figure 5: Mean and standard deviation of average nominal wages by forms of property in regions

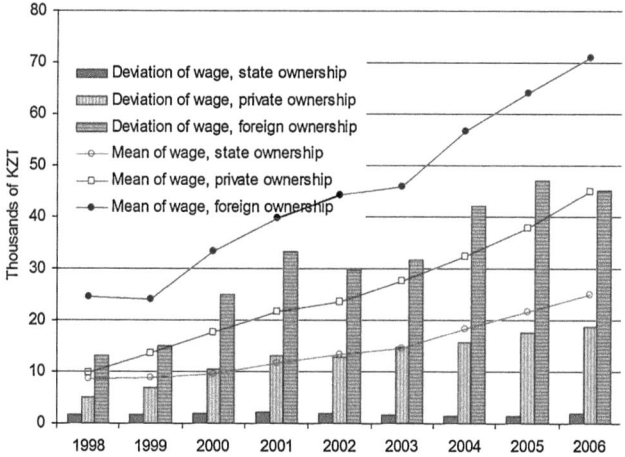

Migration and Remittance Flows within and from Selected Countries of Central Asia and the South Caucasus

Abdul Ghaffar Mughal[18]

1. INTRODUCTION

In the wake of the transformational recession combined with internal conflicts and growing economic disparities in post-Soviet republics, Russia became a magnet for labor from the Ukraine, Moldova, Central Asia, and the Caucasus. The unanticipated exodus from the south to the north and from the west to the east within the former Soviet Union (FSU) created a new corridor in the global migration system, becoming almost a structural feature of the socio-economic landscape of the FSU. Russia is now the second largest immigrant receiving country in the world.

While Russia remains the center of gravity for migration within the FSU, Kazakhstan also turned into a destination for labor from the poorer CIS countries, in particular, the neighboring countries of Central Asia. This paper identifies the main patterns of migration and remittance flows in the CIS-Russia-Kazakhstan corridor. It draws mainly upon the surveys of remittances and poverty carried out in 2007 in Armenia, Azerbaijan, Kyrgyz Republic, and Tajikistan.

The first section examines migration patterns, and the second section discusses remittance patterns. Since the household survey of remittances and poverty was limited to the four countries surveyed – Armenia, Azerbaijan, Kyrgyzstan, Tajikistan (henceforth, CIS-4) –, most of the discussion is carried out from the vantage point of these countries even though the findings may be generalized to other migrant sending countries of the CIS corridor. When the discussion includes the host countries (i.e., Kazakhstan and Russia), we refer to them collectively as CIS-6. When we discuss the results of the survey of remittance senders conducted in Kazakhstan, we include Uzbekistan among the migrant sending countries, as the survey included only remittance senders to these five countries (henceforth, CIS-5).

---

18 The author is indebted to the Asian Development Bank (ADB) for the use of the data from the various surveys in the region conducted in 2006.

2. PATTERNS OF CONTEMPORARY MIGRATION FLOWS IN CIS-4

One of the defining characteristics of the migrant sending countries under consideration is their landlockedness. While under conducive conditions for exports of goods, *geographic landlockedness* need not translate into *economic landlockedness*, the CIS-4 have witnessed serious structural constraints in expanding trade relationships with the outside world and diversifying their economic structure. The structural constraints have acquired added significance in case of Tajikistan.[19]

2.1 Prevalence of Migration in CIS-4

Table 1 gives the share of different categories of migrant households in the country as a whole and by localities within the country. The following similarities across countries are worth noting. First, there is a significant number of households with at least one migrant in the sample in each country. The proportion ranges from a low of 11.18% in Azerbaijan to a high of 37.3% in Tajikistan, with the Kyrgyz Republic and Armenia lying in between the two extremes, with 15% and 18% households with at least one migrant each respectively. Second, there are about three to four times as many households with one migrant member as with two migrant members.

Tajikistan has the highest percentage of households in each category of migrant households compared with any other country in the CIS-4 group. It is not surprising that resource rich Azerbaijan has the lowest proportion of migrant households with at least one migrant.

However, the vast difference between the Kyrgyz Republic and Tajikistan in terms of the incidence of migration is surprising and requires careful analysis. First, while both Tajikistan and the Kyrgyz Republic share many structural constraints (landlockdness, high degree of structural unemployment, limited size of the market, etc.), the severity with which these constraints are faced by these two countries varies considerably. Tajikistan faces unusually severe obstacles in expanding market size through trade with neighbors to the South, North and the East. Second, to some extent, the aftereffects of the civil war that destroyed a significant part of the infrastructure of the country linger over. Last, but not least, the opportunity cost of migration for Kyrgyz farmers is much higher (especially on the North, where land plots are larger) than for Tajik farmers. The comprehensive land reforms implemented in the Kyrgyz Republic early on

---

19 See Gaël Raballand (2003), for a theoretical discussion of landlockedness, and Mughal (2007) for an application to Tajikistan.

provided practically every farmer with some piece of land with ownership rights, boosting agricultural productivity. The failure of land reforms in Tajikistan, and the consequent pressure on farmers to grow cotton in extremely unfavorable economic conditions have kept agricultural productivity low and induced migration as a substitute means of earning livelihood. Thus, for rural households in the predominantly agrarian Tajikistan, too often households find that they can improve their welfare by migrating abroad than by staying home even considering all migration-associated economic and social costs.

Table 1: Migrant Households with more than Two Migrants
by per capita income quintiles of households
(in % of the total number of sample households in country/locality/quintile)

	ARM		AZE		KYR		TAJ	
	One Migrant HHM	Two Migrant HHMs	One Migrant HHM	Two Migrant HHMs	One Migrant HHM	Two Migrant HHMs	One Migrant HHM	Two Migrant HHMs
Country	12.3	4.0	7.8	2.8	10.8	3.0	26.7	8.6
Capital	8.7	3.2	5.5	2.8	5.1	1.3	14.1	2.9
Other Urban Areas	14.8	4.3	7.0	2.2	12.4	2.6	26.8	6.0
Rural Areas	13.9	4.6	9.7	3.2	12.0	3.5	29.2	10.6
	>2 Mig HMs	at least 1 Mig HHM	>2 Mig HMs	at least 1 Mig HHM	>2 Mig HMs	at least 1 Mig HHM	>2 Mig HMs	at least 1 Mig HHM
Country	1.6	18	0.59	11.18	1.1	14.9	2	37.3
Capital	2.1	14.1	0.38	8.67	0.5	7	0.2	17.1
Other Urban Areas	1.6	20.8	0.99	10.18	1.6	16.6	1.4	34.2
Rural Areas	1	19.6	0.46	13.33	1.2	16.7	2.5	42.3

Source: author.

2.2 Determinants of Outward Migration from CIS-4 and the Main Destination Countries[20]

While Russia is the main destination country for migrants from all CIS-4 countries, the degree of dependence on the Russian labor market varies across the four countries: a significant proportion of migrants from Armenia and Kyrgyzstan have found work in high-income OCED countries. While

---

20 This section draws heavily upon the author's previous work (Mughal 2007).

Kazakhstan is the second most important destination of migrants (proxied by remittance senders) from the Kyrgyz Republic, it occupies a distant second place for migrants from Tajikistan and Azerbaijan, as less than 1% of all categories of migrants who send money to these two countries are located in Kazakhstan. The United States remains the second most important country of destination for migrants from Armenia, accounting for about one in four external migrants who send money to Armenia. In contrast, only a small percentage of Tajik migrants lives and work in any country other than Russia.

What are the determinants of outward Migration from CIS-4, and, why do migrants from these countries 'choose' the destinations that they do? The following discussion from a theoretical perspective sheds light on these questions.

2.2.1 The Power of Structural Explanations: The conspiracy of geography and history

Landlockedness

Both nature and history seem to have conspired to make the CIS-4 countries good candidates for the export of labor. A common characteristic that sets the CIS-4 apart is their landlockedness. The nearest seaport to Tajikistan is 2000 kilometers away and 93% of the land consists of non-arable mountains. High mountains separate Tajikistan from Kyrgyzstan in the North and the economic powerhouse in the East, i.e. China. In the presence of serious handicaps in expanding the size of the market through exports, emigration has emerged as a substitute for the export of goods (Raballand 2003; Mughal 2007).

Although other major labor exporting countries in the Russia-CIS corridor, like Moldova, Kyrgyzstan and Armenia, are also landlocked, Tajikistan is unique in many respects. Tajikistan is a no-transit 'landlocked' country whereas Kyrgyzstan can be viewed as a transit country (for the export of goods, say, from China to other Central Asian countries). This is a very important distinction (Raballand 2003).

Lack of Diversification: Continuation of the core-periphery relationship with Russia through labor export for remittances

The demise of the Soviet Union did not fundamentally alter the core–periphery relationship that Tajikistan always had vis-à-vis Russia. One manifestation of a lack of diversification is the historically inherited and government backed cotton mono-culture in Tajikistan which has involved a considerable misallocation of resources and has limited agricultural productivity throughout the post-Soviet period.

Landlockedness coupled with other historical factors go far to explain the heavy dependence of CIS-4 on the Russian labor market. Nowhere is this more obvious than in Tajikistan. Russia and the CIS-4 countries continue their symbiotic relationship today as they did during the Soviet era – albeit the form of mutual dependence has transformed from raw material for subsidies to migrant labor for remittances.[21]

Transformational Crises and Ethno-national Conflicts

Most of the migration from the CIS-4 in the immediate aftermath of the dissolution of the FSU represented a permanent return migration of mainly Russian, German, and other non-Titular ethnic groups to their native homelands. In the case of Tajikistan, the Civil War (1992–1997) added an additional factor that resulted in near total brain drain from the country as the intelligentsia had the most to lose amidst the chaos of the civil war. In addition to Tajikistan, the conflict over Nagorno Karabakh between Armenia and Azerbaijan also created a wave of refugees from these countries.

The first wave of ethno-political migration of a permanent nature that attended upon the demise of the FSU gradually passed into a second wave of temporary and seasonal migration of an explicitly economic character as much of the Diaspora migration ran its course and security risks diminished. Although there is usually a deep undercurrent of economics beneath the apparently non-economic moves, the proximate causes for migration have changed over the period from being predominantly ethno-political to economic in nature. The contemporary wave of migration within the FSU space reflects perceived expectations about differences in earnings and the quality of life. The change in the prime motives go far to explain the change in the character of migration flows from CIS-4: from permanent Diaspora style return migration of non-titular minorities to temporary and seasonal labor migration of the members of the titular majority.[22]

---

21 The author admits to the oversimplification of the relationship between the center Moscow and the Tajik periphery implied by the literary license exercised here, yet it brings the theme of the report into sharper focus.
22 Migrants can be classified as different types, depending on the reasons for migration, including, among others, permanent long-term migrants, temporary migrants, shuttle migrants, transit migrants, asylum seekers, and inter-firm corporate personnel. See Olimova and Bosc (2003) for a detailed discussion of different types of Tajik migrants.

The Easterlin Hypothesis and the Demographic Imbalance in the CIS

While Azerbaijan, Kyrgyz Republic, and Tajikistan have a very high fertility rate, and as a result have a large proportion of young people in the total population, Russia has a negative population growth rate with an aging population.[23] The demographic factor in explaining migration is well documented in the literature.[24] The demographic imbalance between CIS-4 and Russia is clearly visible in the population pyramids of these countries.

Russian fertility rate has been declining for decades; divorce rates have gone up by one third at the same time that marriage rate has gone down by the same percentage (Lissovlik 2005). To quote a recent World Bank report on population aging in the FSU: "The most striking case is the Russian Federation, where the population – which fell from 149 million in 1990 to 143 million in 2005 – is projected to fall to 111 million by 2050" (World Bank 2007a, 45). Nowhere is the demographic imbalance as sharp as between Tajikistan and Russia. Tajikistan is projected to have a growing population at least for the next two decades. Between 2000 and 2005, Tajikistan's population is projected to have grown by 42%, the highest percentage growth rate for any country in Eastern Europe and Former Soviet Union (ibid, 7).[25]

Vent for Surplus Labor

Myint's concept of 'vent for surplus labor' offers additional insights into the migration flows between CIS-4 and the labor receiving countries of Russia and Kazakhstan.[26] The concept of surplus labor is a relative concept. High fertility rates in Kyrgyz Republic, Azerbaijan, and Tajikistan create a labor surplus economy, ceteris paribus, and the well-known Russian demographic crisis makes Russia a labor scarce economy (given Russia's vast land and natural resources).

---

23 Russian policy makers and scholars are keenly aware of the demographic crisis. Russian government announced in July 2006 that it would invite as many as 1 million immigrants from the FSU by offering them citizenship and other benefits, particularly to those who arewilling to settle in the underpopulated regions, such as Siberia.
24 In his famous article of 1961, Easterlin had presented his thesis about the main driving force behind the Atlantic migration from Europe to North America: Europe had a burgeoning population and a very high labor to land ratio.
25 This is despite the potentially lower birth rate thanks to migration to Russia and elsewhere which keeps partners apart for years in many cases.
26 See Myint (1958). If unemployment and underemployment rates are extremely high and themarginal productivity of labor in the emigration country is zero, vent for surplus implies gaining something for nothing.

A high population level or population growth rate per se is not as significant a push factor as the relative proportion of labor to land. Both Kyrgyzstan and Tajikistan may be considered to be labor surplus economies in the sense that only 7% of the total land is arable which makes the labor to land ratio very high. A similar conclusion could be drawn for Armenia.

2.2.2 The Power of the Neoclassical Framework: Widening Income Gap and the Recent Oil Boom

At the time of independence, the per capita income (PCI) differential between Russia and the selected countries of the region was small. The gap between the CIS-4 and the receiving countries has been widening. Although oil has been lubricating the Russian and the Kazakh economies for quite some time, the unprecedented oil boom in Russia and Kazakhstan in the last few years has boosted demand for labor dramatically, especially in certain sectors, such as, construction, oil and gas, mining, and restaurants and catering, and domestic services.[27]

2.2.3 Relative Deprivation and Persistence of Migration

Migration is frequently location specific. A powerful explanation of continued migration from the same community is offered by the theory of relative deprivation (Stark 1991). According to this theory, household utility is a function not only of absolute income but also of income position vis-à-vis other households in the village or reference group. When some households from a community undertake migration and are able to improve their economic status vis-à-vis other households in the same community, the non-migrant households will find themselves relatively deprived, inducing them to migrate as well to keep up with the Joneses.

It must be kept in mind that the type of migration is extremely important while considering the determinants of migration. In fact, the external migrant may differ significantly from a family migrant – a typical external migrant is older, more highly educated, and is highly less likely to be a seasonal and temporary migration.

---

[27] The Putin government is keenly aware of this and has introduced significant subsidies to encourage fertility. Public pronouncements notwithstanding, the Russian government understands that immigration is necessary in the foreseeable future. The Russian government is particularly concerned about areas other than Moscow and St. Petersburg which attract immigrants from abroad as well as from other regions of Russia. Thus, recently, the Russian government decided to invite 10,000 immigrants to the oil rich region of Krasnoyarsk to help develop the new Vankor oil and gas field.

2.2.4 The Emigration Multiplier à la Cumulative Causation

Most empirical studies find that network factors eventually increase in importance over time. A self-feeding momentum gets built into migration. In a process that has been termed "cumulative causation," each time a migrant leaves, he "alters the social context within which subsequent migration decisions are made, typically in ways that make additional movement more likely" (Myrdal 1957; Massey *et al.* 1993, 451). The explanation for snowballing migration is that migration "tends to increase in prevalence and become more diverse because transnational movement causes relatively permanent changes in individual motivations, social structures, and cultural milieus, and these changes cumulate over time to change the context within which subsequent migration decisions are made. As more people are induced to migrate, knowledge and network connections expand further, inducing more people to migrate, and so on" (Massey *et al.* 1994, 1528).

Additionally, the theory of 'cumulative causation' of migration implies that once a critical threshold of immigrants arises in the host countries, migration can become a self-reinforcing phenomenon mainly because of reduction in transaction costs. Information flows and a whole network of support make it easier for prospective migrants to try their luck. Newcomers can find a place to live and relatives and friends can assist in job location. Qualitative research confirms the findings in other high-emigration countries.

Within the emigration area, there may be secondary effects, including changes in the distribution of income, changes in the distribution and usage of farm land, and changes in consumption habits. These secondary effects in turn may independently induce additional migration.

In accordance with the predictions of the cumulation theory of migration, with the buildup of networks and critical thresholds of Tajik migrants in many host cities in Russia, migration has acquired a logic of its own. It functions very much like the Keynesian multiplier: an initial dose of emigration reaching a critical threshold level leads to a multiple expansion of emigration. A relatively low cost of transportation and a visa free regime with Russia greatly facilitates the process of migration.

The cumulation theory of migration implies that migration has a multiplier effect once it reaches a critical threshold. Thus, once colonies of migrants are established abroad, relatives and friends back home find it easier to join the migrants in the host country – usually in the same city, and, at least temporarily on arrival, in the same living quarters – thanks to the informational, logistical, and financial support they provide.

A high positive correlation between the number of acquaintances abroad and the number of migrant households in each country can serve as an indirect test of the network theory or the theory of cumulative causation of migration although causation could run both ways. Tajik households stand out in terms of having the largest number of acquaintances abroad that can potentially help in finding work abroad. Tajik non-receiving households also have a large number of acquaintances on average compared with households in other three countries. This shows the overwhelming importance of network for current or future cohorts of Tajikistan migrants. Although Russia experienced windfall oil profits in 2000–2004 and Russian GDP has been growing at an average rate of 6% since 1999, we cannot attribute the dramatic shift in migration solely to the oil boom.

The share of migrants in Russia coming from Central Asia and South Caucasus has been steadily increasing for two other main reasons: need for replacement of migrants from other Slavic republics, and, emigration of highly skilled Russians to high income countries of OECD. First, during the Soviet era, immigration from the Slavic Republics (Ukraine and Belarus) used to compensate for labor shortages in Russia. Emigrants from these countries have now been finding more lucrative alternatives in high-income EU countries; thus creating job opportunities in Russia for migrants from poorer regions of the CIS. Second, to add to the labor shortage, native Russian labor force – which was already in short supply – has also been moving to the high-income countries of Western Europe, Israel, and the United States (Heleniak 2002).

A related observation is that migrants from the CIS-4 countries have found niches in host countries that the native workforce finds least attractive at the going rates. The exodus of ethnic Russians, Germans, and other non-titular groups has vacated employment positions that have been filled by native labor or in some cases by imported labor from other rich countries. The most unpleasant and dead-end jobs shunned by natives and highly skilled immigrants from the rich countries have been reserved for migrants from the poorer CIS countries, mainly the Central Asian neighbors. The segmentation theory of displacement offers useful insights in comprehending this phenomenon. Proponents of this theory hold that low skilled immigrants, particularly the undocumented workers, are offered jobs (such as cleaning or picking strawberries) that native workers shun (Piore 1979).

Table 2: Determinants of Migration to Kazakhstan and Russia from CIS-4

	Per-Capita Income	Percent of Kazakhstan	Percent of Russia	Wages as % of Kazakhstan	Wages in % of Russia	Annual employment growth	Annual population growth	Real GDP in % of 1989
	2005	2005	2005	2005	2005	2000/5	2000/5	2006
Armenia	4,428	63.9	46.2	31.8	37.7	-3	-0.4	130.6
Azerbaijan	4,374	63.1	45.6	34.9	41.4	0.8	0.8	129
Kyrgyzstan	1,695	24.5	17.7	17.5	20.8	1.8	0.9	87
Tajikistan	1,134	16.4	11.8	8.1	9.6	3.9	2.1	60.5

Note: Per capita income is based upon 2000 PPP.
Source: Author's calculation from UNECE.

Persistent and growing gaps in expected earnings, coupled with differential rates of population growth in recent years go far to explain the attraction of the Russian and the Kazakh labor markets for the workers from CIS-4. This is clearly shown in Table 2. The difference between the per capita income in Tajikistan and the per capita income in Russia as well as Kazakhstan is the greatest, with the Russian per capita income being ten times the per capita income in Tajikistan. A visa-free regime between CIS-4 and the two host countries also facilitates migration by reducing transactions cost.

2.3 International Migration into Kazakhstan

Within the CIS corridor, the main vector of inter-state migration follows the same principle: from labor-surplus / low expected earnings[28] / low standard of living countries to labor-deficit / high expected earnings / high standard of living countries. Thus, along with Russia, Kazakhstan has become a major pole of attraction for seasonal labor migrants from poorer Central Asian countries, reflecting Kazakhstan's growing prosperity, demographic deficit, and relative economic liberalism. The oil boom starting 1999 gave labor migration from the CIS-4 into Russia, and increasingly into Kazakhstan, a shot in the arm.

---

28 As formulated in the seminal paper of Todaro (1976), expected earnings involve potential earnings adjusted by the probability of getting a job.

2.3.1 Size, Nature, Origin, and Sectoral Composition of Migrant Labor in Kazakhstan

Immigration in Kazakhstan clearly shows signs of a secular increase starting 1997–98. In fact, while the graph of emigration is marked by secular decline after 1994–95, the graph of immigration is increasingly U-shaped. In the last few years after 2004, the immigrant flows have reportedly offset the emigrant flows in Kazakhstan.

The growing labor deficit in Kazakhstan reflects excessive labor demand in the face of sizeable reduction in the labor force since the breakup of the FSU. Of all the ex-Soviet republics, Kazakhstan had the largest proportion of Russians in the population. During the wave of diaspora/return migration from 1989 through 2004, it lost 20.7% of its population due to mass emigration of not only ethnic Russians, but also Germans, Jews, and other groups who were forcibly settled in Kazakhstan during the Stalin period. Over 800,000 Germans left Kazakhstan soon after the breakup of the FSU.

Realizing its growing demographic deficit and the political significance of having a sizeable Kazakh majority, the Kazakh government launched its own program to replenish the population through repatriation of the people of Kazakh ethnicity that have been living for centuries in many countries outside Kazakhstan. Thus, from 1991 to 2001, 183,652 Kazakhs returned to Kazakhstan (IOM 2002, p. 16), although there are still an estimated 5 million Kazakhs residing outside Kazakhstan, mainly in China (up to one million), Mongolia, Uzbekistan, Afghanistan, Iran and Russia. The success of the repatriation program has been limited partly by Kazakh policy, which prohibits dual citizenship.

The share of Kazakhstan in the regional labor market has been increasing due to its closer location, less overt hostility towards migrants of Tajik, Uzbek, and Kyrgyz origin than in Russia, a more similar climate, and cultural/religious/linguistic affinity between the Kazakhs and the Kyrgyz, and, to a lesser degree between Kazakhs and Tajiks, and between Kazakhs, Uzbek and Azeries.[29]

Based on a survey conducted by the World Bank in 2005, the overall stock of immigrants in Kazakhstan was estimated to be about 2.5 million people or 16.9% of the total population with top 10 source countries being Russia, Ukraine, Uzbekistan, Germany, Belarus, Azerbaijan, Turkey, Poland, Tajikistan,

---

29 Kazakh, Kyrgyz, Uzbek, and Azeri languages are all Turkic based.

and Moldova. Russia is both an exporter as well as an importer of labor from Kazakhstan. Foreign labor comprises a significant proportion of the labor force in Kazakhstan.

The estimates of labor migrants from CIS are imprecise because the bulk of migrants is irregular/undocumented. According to a recent World Bank (2007a) report, there were about three million migrants in Kazakhstan in 2002. The same report also estimated that 9.9% of all these migrants in Kazakhstan were 'irregular' (p. 152). A visa free regime coupled with prohibitive legislation concerning employment of foreign labor, and high payroll taxes go far to explain the preponderance of undocumented labor from CIS. Published figures of registered legal migrants from the Statistics Agency of Kazakhstan show that only tiny proportion of the migration labor is officially registered. What percentage of the total migration pool in Kazakhstan is accounted for by legal migration? The number of registered migrants as of the first half of 2007 was to be 171,600, which leaves over 85% of the estimated 1,171,600 migrants in 2006–7 in the pool of the undocumented (ADB 2008).

The largest part of migrants, especially low-skilled workers and those employed in small and mid-sized projects continue to be hired illegally.[30]

While economic growth in Kazakhstan is keeping rapid pace at above 10% on average over the past eight years, a growing deficit of low skilled workers is especially marked in certain non-tradable sectors. Most of the migrants work in non-tradable sectors of the economy. As in Russia, the construction boom in residential, commercial, and civil engineering (infrastructure) projects in Kazakhstan contributed to the spectacular growth in the non-oil sectors. The amnesty granted to irregular migrants (first of its kind in the region) during August through December 2006 revealed that a vast majority of legalized migrants (115,072 or 70%) worked in construction and 22,998 (14%) in the services sector. The amnesty also confirmed what earlier reports had shown: among tradable sectors, agriculture leads with 13,258 (8%) of total migrant workers (including lucrative tobacco and cotton production). The number of

---

30 Illegal entry and/or illegal presence in the host countries are all pervasive. The form of illegality varies from country to country depending on the incentives built into the legal system. The definition of 'illegality' of migration is very ambiguous in the CIS context. A political economy analysis reveals that there are built-in incentives in the system to perpetuate undocumented migrants. Paradoxically, illegality often arises from lack of legal channels for registration. The visa-free regime between the host and home countries within the CIS means that almost all migrants enter the host countries legally; yet illegality stems from the further requirements of different forms of registration, which vary from republic to republic. For example, most Central Asian countries and Russia require foreigners to register with the authority within a certain period, after which an immigrant is considered 'illegal'. Yet this is often difficult or impossible for migrants to do.

migrants from the region in the tobacco and cotton plantations had been growing steadily from 1998-2000 (IOM 2002, 93).

2.3.2 Origin of Migrants into Kazakhstan

While Kazakhstan imports skilled labor from Russia (to the oil industry, transports, and construction) and countries from outside CIS, Central Asian neighbors have become the main suppliers of unskilled labor to Kazakhstan. A special survey in Kazakhstan was conducted to study the phenomenon of migration from the vantage point of the remittances receiving countries.

The correspondence between the citizenship and ethnicity of remittance senders is less than 100%. For instance, a significant proportion (13.5%) of remittance senders reported being Kazakh citizens even though people of Kazakh ethnicity among the remittance senders constitute only 8.7%. As expected, citizens of the Kyrgyz Republic and people of Kyrgyz ethnicity dominate the remittance senders (both about 23%). There are more people of Armenian, Azeri, Tajik ethnicity than citizens from these countries. Uzbekistan is an exception in having approximately 36% of its citizens represented among the remittance senders but less than 32% people declaring their ethnicity to be Uzbek. We do not have enough information from the surveys to precisely identify the ethnicity of Uzbek citizens of non-Uzbek origin. The surveys are likely to have under-represented Uzbek citizens who are of non-Uzbek origin.

3. REMITTANCE FLOWS IN SELECTED COUNTRIES OF CENTRAL ASIA AND THE CAUCASUS

There is a twin mirror relationship between migration and remittances. As people move, so does money. International labor movements and workers' remittances have long been a structural feature of the global economy. Following the breakup of the FSU, the countries of CIS-4 began to witness cross-border monetary flows attending upon Diaspora/return migration, which in many cases created transnational families. The cross border monetary flows became massive when the initial wave of permanent migration passed into the second wave of temporary and seasonal migration dominated by economic considerations. Although remittances have been a part of the cross border flows in the CIS, only recently did it attract attention from political and policy making circles. Remittances came into the limelight mainly due to two main reasons in the Kyrgyz Republic and Tajikistan: (i) a rapid increase in migration of workers to the Russian Federation (and, in case of the Kyrgyz Republic, increasingly to Kazakhstan) thanks mainly to the rising demand for labor in the wake of the oil boom in these countries; and (ii) macroeconomic stabilization and relaxation of foreign exchange controls. The unprecedented and growing appearance of formal remittances as well as 'errors and omissions' in the BOP attending upon the high levels of informal transfers by shuttle traders caught the attention of policy makers and international financial institutions.

The significance of remittances was recognized in Armenia earlier than in any other country in the CIS-4 group. The Armenian Central Bank has been collecting data on transactions carried out by money transfer operators (MTOs) since 2001. Remittances in the BOP of Armenia are calculated based on the formal remittance transactions made through the banking sector and money transfers through non-bank organizations as reported to the CBA, as well as the NSS (Integrated Living Conditions Survey) conducted each year which helps to categorize the amounts into different categories and incorporate other elements such as data on border workers. Little attention was paid to the remittance flows in Azerbaijan until fairly recently.

3.1 Measuring Remittances in Central Asia and the South Caucasus

3.1.1 Remittances in the Official Balance of Payments

According to the International Monetary Fund's (IMF's) Balance of Payments Manual (1993), remittances consist of three components:

- *Compensation of employees (COE)* falls under the income account (of the current account) which includes income earned abroad by seasonal or short-term workers (foreign residents for less than a year).

- *Workers temporarily working abroad* are still considered as domestic residents and thus their wages earned in the foreign country represent a payment from a foreign resident to a domestic one.

- *Workers' remittances cover current transfers* (under the current account of the BOP) by migrants employed in new economies where they are considered residents. A migrant is a person who comes to an economy and stays there, or is expected to stay for a year or more.

- *Migrants' transfers* are contra-entries to the provision of a resource (such as grants and gifts in kind or financial form) that arise from the migration of individuals from one economy to another without a quid pro quo. Migrants' transfers fall under the capital transfers account (of the capital account). This includes the transfer of financial assets when migrants change residence.

Monitoring of remittances in CIS-4 has a very short history, mostly non-existent for the first decade following the breakup of the FSU. The central banks of the countries of Central Asia and South Caucasus have limited capacity to monitor these monetary flows, with the possible exception of Kazakhstan and Armenia.[31] The bulk of recorded remittances consist of workers remittances and compensation of employees. Considering only these two items, the data for the four destination and two source countries in the CIS corridor is published in the IMF yearbooks on BOP from 1998 through 2006 and are presented below in Table 3.

Table 3: Formal Remittances Inflow/Outflow (in millions of US $)

	1998	1999	2000	2001	2002	2003	2004	2005	2006
Armenia	1667	183	182	209	264	336	638	765	1361
Azerbaijan	83	71	69	110.2	168	164	212	637	807
Kyrgyz Republic			8	8	37	78	189	322	482
Tajikistan				20	124	160	278	467	1119
Kazakhstan (gross outflows)	471	356	440	487.3	595	802	1354	2000	3036
Russia (outflow)								6,989	11,438

Note: World Development Indicators are based upon data collected in IMF Balance of Payments Statistics Yearbook.
Source: WB; WDI (2007).

First, the figures are marked by discontinuous increases. The only anomaly present here is the U-shape pattern of transmission in Azerbaijan where reported formal remittances fell from 1998 through 2000 before showing a secular increase for the subsequent years as for other countries. These discontinuities in reported formal remittances are present in all countries.

Second, there is wide variation in the proportion of migrant transfers among the four remittances destination countries. Thus, although in relative terms, the growth is extremely sharp for Kyrgyzstan, Armenia, and Tajikistan, the increases in absolute terms during these years are particularly dramatic for Tajikistan and Armenia where the amount of remittances increased from $467m to $1119m and from $765 to $1360m respectively. Also, while migrant transfers

---

[31] There is already substantial evidence of an increasing knowledge divide among various countries of Central Asia. See Mughal (2007, 167), McGlinchey (2006), and Milam and Vanjani (2008).

are almost non-existent in the Tajik balance of payment, they constituted over 10% of the total remittances flows in Azerbaijan in 2005 (IMF 2005).

Net errors and omissions in the balance of payment of the two Central Asian countries, mainly Kyrgyzstan, reflect the presence of sizeable informal foreign currency transactions. Net errors and omissions, in essence, is a balance of all in the economy. These informal transactions include imports and exports carried out by shuttle traders, smuggling of some commodities such as gasoline from Uzbekistan and Kazakhstan, where prices for these commodities were and still are subsidized; other receipts from illegal activities; and, finally informal transfers of workers abroad. Of all the CIS-4 countries, Kyrgyz Republic has registered the sharpest increase in net errors and omissions since 2003, indicating a growing inflow of foreign currency to Kyrgyzstan.[32] Of course, as to which components of informal flows contributed to this growth is unknown, but it is highly likely that some part of this growth is related to informal transfers.

Not only has the level of remittances increased since 2002, but also the rate of growth of remittances has increased in most of the receiving countries.

The growth pattern of formal remittances mainly reflects increasing demand for migrant labor attending upon the oil boom in the main destination country, Russia. However, this factor alone does not explain the dramatic jump in formal remittances. All over the world, formal remittances have shown a remarkable elasticity of demand: with improvement in the technology of transfer and significant reduction in transfer fees resulting from fierce competition in the money transfer business, many migrants have shifted from informal to formal means of transfer. Certain factors specific to each country also contributed to the shift in the mode of transfer. In case of Tajikistan, abolition of the 30% fee on remittance receipts has been an important factor in boosting transfer through formal banking channels.

3.2 Patterns of Remittances as Revealed by the Household Surveys in CIS-4

Table 4 below provides cross tabulation of migrant and non-migrant households by remittance receiving status. The table shows that the proportion of remittance receiving households among the migrant households is significant in each country, and, underscores the twin mirror relationship between the two flows.

---

32 Based on communication with the authorities of Kyrgyz National Bank as well as the country report on remittances and the financial sector by ADB (2008b).

Migration in Central Asia and the South Caucasus

Table 4: Migrant households with more than two migrants per capita income quintiles of households

(in % of the total number of sample households in quintile)

	ARM		AZE		KYR		TAJ	
	HHs with One Migrant HHM	HHs with Two Migrant HHMs	HHs with One Migrant HHM	HHs with Two Migrant HHMs	HHs with One Migrant HHM	HHs with Two Migrant HHMs	HHs with One Migrant HHM	HHs with Two Migrant HHMs
Lowest Quintile	6.9	4.0	1.7	1.7	6.6	0.9	14.8	3.8
Second Quintile	7.5	3.3	4.2	1.2	9.6	1.8	20.9	6.4
Middle Quintile	11.5	4.0	7.3	2.6	10.0	2.9	26.4	7.7
Fourth Quintile	13.5	3.2	8.8	2.6	15.2	2.8	34.1	10.0
Highest Quintile	22.2	5.7	17.1	5.9	12.6	6.5	37.4	15.1
	>2 Mig HMs	at least 1 Mig HHM	>2 Mig HMs	at least 1 Mig HHM	>2 Mig HMs	at least 1 Mig HHM	>2 Mig HMs	at least 1 Mig HHM
Lowest Quintile	1.5	12.4	0.26	3.59	0.6	8.1	0.9	19.5
Second Quintile	2	12.9	0.26	5.64	0.1	11.5	2	29.2
Middle Quintile	1.2	16.7	0	9.88	0.5	13.4	1.8	35.9
Fourth Quintile	0.8	17.6	0.9	12.29	0.9	18.8	1.4	45.4
Highest Quintile	2.3	30.3	1.54	24.49	3.5	22.7	3.8	56.4

Notes: Quintiles based on household income including remittances; the household categories are exclusive.
Source: author.

The variation in the proportion of remittance receiving households in the four countries reflects the differences in the incidence of migration among the households in the four countries. Thus, proportion of remittance receiving households ranges from the low of 12.7% in resource-rich Azerbaijan to 40% in Tajikistan, with the Kyrgyz Republic and Armenia lying in between the two extremes with 15.6 and 26.3% each respectively. Again, Tajikistan occupies the first place with 96.3 % of the migrant households receiving remittances from migrant members in 2006. Both Armenia and the Kyrgyz Republic seem to have similar proportions of remittance receiving migrant households, i.e., 87% and 86% respectively.

Just as not all households with migrants receive remittances, not all migrants send remittances to their households. Thus, the proportion of migrants that send remittances need not be equal to the proportion of migrant households that receive remittances. For instance, in the Kyrgyz Republic, while 79% migrants are reported to have sent money in 2006, 86% of the migrant households reported receiving remittances. The percentage of households receiving remittances among migrant households in Tajikistan is 97 whereas the percentage of migrants who do send money is much lower.

Paradoxically, the percentage of migrant households that receive remittances is higher than the percentage of migrants who send money. This apparent paradox is resolved if one considers that the decision to migrate and to remit is a household decision. Thus, some households may decide to send more than one family member abroad. The latter may depend crucially on household size, which varies considerably among CIS-4 countries (Table 4). Thus, when a household sends more than one migrant, not receiving money from a particular household member does not mean no money was sent by that migrant. It could simply mean a division of labor among migrant household members where one person takes charge of collecting and sending money or sends all that he earns and the other takes care of day-to-day expenses for both. This explains why, in general, the percentage of remitting migrants is lower than the percentage of migrant households that receive remittances.

Significance of Remittances: Remittances and Income

Table 4 above gives the share of different categories of migrant households in the sample by income quintiles. The following similarities across countries are worth noting. First, there appears to be a linear trend in terms of the percentage of households represented in each successive income quintile. Second, regardless of the number of migrants per households, all countries have a higher percentage of households represented in the highest income quintile than in any other income quintile (with the exception of households with one migrant member in the Kyrgyz Republic).

With these similarities, there are some important inter-country differences in terms of the proportion of remittance receiving households in each income quintile. Tajikistan has the highest percentage of households in each quintile compared with any other country in the CIS-4 group with the proportion of migrant households in the highest income quintile reaching over 56%!

Another difference worth noting is that the linear relationship between incidence of migration and income quintile is weaker in case of Armenia and Kyrgyz Republic among households that have one migrant. There is no clear trend for this category of households in Armenia, and, in Kyrgyz Republic, the fourth income quintile has over 15% of the households with one migrant compared with the fifth income quintile, which has 12.6% of the households with one migrant in the total sample of households. If we exclude remittances from the income of HH and make a breakdown by quintiles, we can observe a reverse trend – i.e. the number of remittance receiving HH increases from the richest quintiles to the poorest one (by 3.3 times). It should be noted that this situation is typical for the HH that received remittances in cash as well as those that receive them in kind.

While the positive correlation between international migration and income is unmistakably present in all CIS-4 countries, it does not, by itself, imply a causal relationship between international migration and household income. Undoubtedly, the rich households have the greater ability to invest in migration, but their wealth could also be a consequence of international migration. Panel data may help resolve the issue of whether international migration is the cause or the consequence of higher income. However, it should be noted that in the context of the CIS corridor of migration low income is weak a deterrent of non-migration given the proximity of the host countries, a visa free regime, and the presence of strong ethnic networks. Moreover, once we exclude remittances from the income of households, the proportion of remittance receiving households in the poorest quintile increases at the expense of the richest quintile by a factor of 3.3. This suggests a stronger negative correlation between remittance receiving status and poverty.

Regarding the share of remittance receiving households in the total sample of households, several points are worth noting. First, the share of remittance receiving households in the total sample of households is greater than the share of any type of migrant households in the total sample for each country of the CIS-4. This finding can be explained in terms of the presence of external (non-household member) migrants who are not household members yet send money to the households. Another important point to note is the low margin of error as revealed by the narrowness of the 95% confidence interval. Second, in all countries the largest variance in income is observed in the capital city and the lowest is observed in the rural areas (with the possible exception of Armenia where the 95% confidence interval in the capital and the rural areas is about equal. This reflects the greater socio-economic homogeneity of the rural population in these countries relative to the population of the capital cities. There are significant differences among the four countries in the variances of the average remittances inflows, ranging from the lowest of about $85 in Tajikistan to the highest of about $378 in Azerbaijan.

Furthermore, households were asked to rank the flow of remittances in 2006 relative to 2005. Some interesting common patterns emerge. Typically, households in all countries stated that the remittance flow has either remained the same or increased. In all cases except Armenia, respondents typically reported higher remittances. In case of Armenia, half of the households reported no change in the flow. Those who claimed to have received lower amount were in a minority in all countries, ranging from approximately 11% in Kyrgyz Republic to slightly over 28% in Azerbaijan.

Those who reported reduced flow of remittances in 2006 relative to previous years are not equally distributed in different quintiles in the CIS-4 countries. Not surprisingly, households that reported lower amount of remittances in 2006 in the two countries of Central Asia fall disproportionately in the two bottom quintiles; in general, the richest fifth quintile recorded the lowest share of those who declared a reduction in remittances flows except for Azerbaijan. However, there is no clear pattern in the two countries of the Caucuses with Armenia showing a slightly higher proportion in the middle quintile than in the bottom quintile and Azerbaijan showing a higher proportion in the top quintile (32.8%) than in other quintiles except the bottom one (44. %).

In terms of locality, there is no clear pattern. Azerbaijan shows the highest proportion of recipients of lower amount of remittances located in the capital city (53.3%) and Armenia shows the lowest (18%).

The proportion of households who declared increased or the same share of remittances in the household income in 2006 is, in general, greater than the proportion of households who reported increased or same 'amount' of remittances in 2006 compared with 2005. The lowest share (53.7%) of such households is observed in Kyrgyz Republic with Armenia showing the highest share of 84.8%, followed by Tajikistan (81.8%), and Azerbaijan (76%).

Household Income with and without Remittances

Changing from one measure of income to another moves families up or down the income distribution. Examining the movement of families among income quintiles shows how big an effect excluding remittances has on the composition of each quintile. Changing from the family to the household as the unit of analysis combines family incomes in multifamily households, which moves families in those units up the distribution and consequently moves other families down. Families with children are generally larger than childless families, so adjusting for family size drops their incomes more than the incomes of smaller families. Since Tajikistan households are in general larger than non-Tajik households, we should expect an upward bias in depicting the income distribution in Tajikistan in terms of households rather than families and in terms of families rather than individuals

How much difference remittances make to the income of remittance receiving households can be gauged from the fact that when remittances are excluded from the measure of total income, the picture changes drastically. The number of households in the poorest quintiles increases with their simultaneous decrease in the richest quintile (ADB 2007a). This suggests the importance of remittances in fighting poverty: a major shock that results in significant reduction in remittances flow can change people's fortunes dramatically.

3.3 Remittance Transfer Channels

According to the survey of households (HHS) and the survey of returned migrants (RRS) conducted for the study in each of the 4 labor exporting countries, between one-third and two-thirds of migrants, depending on the country of origin, used informal channels—or methods outside of the formal financial system such as bank transfers—to transmit remittances at some point. Paradoxically, the distribution of channels of transmission from Kazakhstan shows preponderance of informal remittances. The apparent inconsistency is resolved once we consider the fact that Kazakhstan still accounts for a very small fraction of all remittances inflows in the four countries; most of the remittances originate from Russia. Since the study was limited to CIS-4, we do not have information on the breakdown of remittances outflows from Russia by channel of transfer from our survey but the growth patterns of formal remittances to these countries can be discerned in the figures regularly published by the Central Bank of Russia (CBR).[33]

The study covered all formal and informal channels available to migrants from each republic for sending remittances. The survey asked questions about the use of three formal and four different informal remittance transfer channels (including "other"). The formal channels include banks, MTOs, and, postal services. Informal channels include courier service, friends, relatives and other people. The preferences of households and migrants on the choice of the transfer channels were also recorded.

---

33 Although Russia attempts to capture remittances sent through informal channels in the balance of payments statistics, the data is not available by country. See World Bank (2005) for the results of a survey of central banks.

Detailed results on transfer channels show some interesting patterns, which warrant careful attention. First, all countries except Azerbaijan show heavy usage of formal channels by migrants, ranging from 70% in Armenia to 81% in Tajikistan[34] Significant parts of Kyrgyzstan are covered by banks and MTOs (ADB 2007a).

Second, the average amount of informal transfers is greater than the average amount of formal remittances received by the remittance-receiving households. This is implied by the observation that the proportion of households that receive remittances through formal channels is greater than the share of formal remittances in the total amount of cash remitted.

Third, not all households rely on a particular mode of transfer exclusively: some households that use formal channels continue to rely on informal channels for some remittances.

Fourth, the use of postal services is negligible in all countries, the highest being 1.3% in Kyrgyzstan. This reflects the relative inefficiency that characterizes postal services across these countries: it is slow, unreliable, and expensive. Its use by some households can be explained by the non-availability of alternative channels in some remote areas. Fifth, among the households that reported receiving money through informal channels, the greatest reliance is placed on migrants carrying the money themselves compared with other methods of informal transfers. However, there are some differences between the countries of Central Asia and South Caucasus: migrants from Central Asia are three to four times more likely to bring the money themselves than entrusting it to friends or relatives. Reliance on friends and relatives reaches almost 30% in Azerbaijan! The use of friends or relatives as couriers is two to four times higher in the South Caucasus. This appears to be a reflection of the higher degree of trust between non-related individuals associated with ethnic entrepreneurship.[35] Such networks appear to be more developed in South Caucasus, and, are associated with the higher degree of entrepreneurship exhibited by migrants from these two countries relative to the migrants from Central Asian countries.

---

34 The 2005 Khatlon Living Standard Survey (KLSS 2005) had shown that 71% of all remittances were transferred through formal means. The results of the ADB survey show 59% of all cash remittances are transferred through formal channels. This apparent discrepancy may signify important regional differences since KLSS 2005 was confined to the Khatlon oblast of Tajikistan and had hardly any migrants to Kazakhstan in the sample.

35 See Light and Karageorgis (2004) for an excellent discussion of immigrant entrepreneurship.

Sixth, reliance on individuals who are neither friends nor relatives is rare in all these countries. Couriers are seldom used except in Armenia where 4% household members reported using courier services for transferring money. These courier services may involve bus drivers or train personnel who may be utilized by some individuals for small amounts.

3.4 Comparison of Remittances Flows in BOP and the Household survey

While, it is relatively easy to calculate remittances sent through formal channels (assuming transfers conform to the conceptual definition of remittances), it should be kept in mind that not all remittances sent through formal channels are properly speaking migration remittances.[36] Estimating the size of informal flows is almost impossible without a well-designed survey of senders or receivers or both (World Bank 2006a, p. 108). Herein lies one of main justifications to carry out a household survey.

Remittances estimated through balance of payments are, in general, greater than remittances estimated directly through household surveys. Estimates based on the official balance of payments in 2006 show a very high volume of remittances in each of the CIS-4 countries: Armenia ($960m), Azerbaijan ($806m), Kyrgyz Republic ($730m), and Tajikistan ($1119m).

While the household surveys conducted in the countries under consideration yield information on the extent of reliance upon informal remittances, which can be utilized to correct for the downward bias in the BOP figures because of the neglect of informal remittances, household survey also has its own well-known bias, i.e., underreporting of income. In fact, the latter bias tends to overwhelm any correction of the former in the remittances data estimated from the household survey.

The household surveys revealed a wide range of differences among the CIS-4 countries in terms of reliance upon formal as opposed to informal channels of transfer, ranging from 23% in Kyrgyzstan to 70% in Azerbaijan, with Armenia and Tajikistan showing the share of informal remittances in the total to be 42 and 41 percent, respectively. The differential share of informal to formal transfers in the four countries affects the adjustments in the estimates significantly.

---

36 For Tajikistan, see Kireyev (2006) and Mughal (2007) for a detailed discussion of the methodological issues involved.

3.5 Source Countries of Remittances

Russia is the main source of remittances for all countries, both in terms of the percentage of all *remittances flows* but also in terms of the percentage of all *remitters* (both household members as well as external migrants). The significance of Russia as the main source country is evident from the percentage of all remitters that are located in Russia, accounting for a minim of 85% of all remittances flows from household members in Armenia in to the maximum of 98% of all flows from household members in Tajikistan. The predominance of Russia is invariant to the type of migrant.

Kazakhstan is the second largest source of remittances for the Central Asian Countries. Kazakhstan is the second largest labor market for migrants from Kyrgyz Republic and Tajikistan, and, this explains why it is also the second largest source of remittances flows into these countries.

There is a twin mirror relationship between migration and remittances. Increasing temporary and seasonal migration to Russia and Kazakhstan is closely associated with increasing remittances flows from these two countries to CIS-4.

The proportion of *household migrant members* who send money to Armenia from the US remains significant but is much lower (4.5%).

REFERENCES

Asian Development Bank (2008), 'Remittances and Poverty in Central Asia and South Caucasus: Country Report on Tajikistan,' *Technical Assistance Consultant's Report – DRAFT.*

Asian Development Bank (2007a), 'Remittances and Poverty in Central Asia and South Caucasus: Country Report on Kazakhstan,' Manila.

Asian Development Bank (2007b), 'Remittances and Poverty in Central Asia and South Caucasus: Country Report on Azerbaijan', *Draft for Discussion at Second Working Meeting*, Kuala Lumpur, Malaysia, 14–16 October 2007.

Asian Development Bank (2007c), 'Remittances and Poverty in Central Asia and South Caucasus: Country Report on Armenia', *Draft for Discussion at the Country Seminar Yerevan*, Armenia 27 November 2007.

Asian Development Bank (2007d), 'Remittances and Poverty in Central Asia and South Caucasus: Country Report on Kyrgyz Republic', Draft for Discussion at the Country Seminar Bishkek, Kyrgyz Republic 22 November 2007.

Anderson, K.H., Pomfret R. and Usseinova, N. (2000), 'Education in Central Asia during the transition to a market economy', in: DeYoung, A.J. and Heyneman S.P. (eds.), *The*

Challenge of Education in Central Asia, Greenwich, CT: Information Age Publishing, pp. 131–152.
Easterlin, R. (1975), 'An Economic Framework for Fertility Analysis', *Studies in Family Planning*, 6, 3, pp. 54–63.
Heleniak, T. (2004), 'Migration of the Russian Diaspora after the Breakup of the Soviet Union,' *Journal of International Affairs*, 57, 2, pp. 99–117.
Heleniak, T. (2002), 'Migration Dilemmas Haunt Post-Soviet Russia', online at: http://www.migrationinformation.org/feature/display.cfm?ID=62, accessed at: 30 March 2012.
IANS, 'Russia in demographic crisis, to support more births', online at: http://in.news.yahoo.com/070101/43/6aqf3.html, accessed at: 16 March 2007.
IFAD (2007), 'Sending Money Home: Worldwide Remittance Flows to Developing and Transition Countries', online at: http://www.ifad.org/remittances/maps/index.htm, online at: 30 March 2012
IMF (2005), 'International Financial Statistics Yearbook', Washington D.C.: IMF.
IOM. (2002), *Migration Trends in Eastern Europe and Central Asia: 2001-2002 Review*, Geneva: IOM.
Kireyev, A. (2006), 'The Macroeconomics of Remittances: The Case of Tajikistan', *IMF Working Paper No. 06/2*.
Light, I. and Karageorgis, S. (1994), 'The Ethnic Economy', in: Smelser, N. and Swedberg, R. (eds.), *Handbook of Economic Sociology*, New York: Russell Sage Foundation, pp. 647–667.
Lissovlik, Y. (2005), 'Opinion: Migration. A Blessing for Russia', online at: http://www.prime-tass.com/news/show.asp?topicid=65&id=377133.
McGlinchey, E. (2005), 'Digital Divide in Central Asia: Comparing ISP Policy', *paper presented at the annual meeting of the Midwest Political Science Association*, Chicago, IL: Palmer House Hilton.
Mughal, A.G. (2007), 'Migration, Remittances, and Living Standards in Tajikistan: A Report Based on Khatlon Remittances and Living Standards Measurement Survey' Dushanbe: IOM.
Myint, H. (1954–1955), 'The Gains from International Trade and the Backward Countries', *Review of Economic Studies*, 22, 2, pp. 129–142.
Myint, H. (1958), 'The 'Classical' Theory of International Trade and the Underdeveloped Countries', *Economic Journal*, 68, 270, pp. 317–337.
Olimova, S. and Bosc, I. (2003), 'Labour Migration from Tajikistan', Dushanbe: IOM.
Piore, M. (1979), *Birds of Passage: Long-Distance Migrants and Industrial Societies*, Cambridge: Cambridge University Press.
Quillin, B., Segni, C., Sirtaine, S. and Skamnelos, I. (2007b), *Remittances in the CIS Countries: A Study of Selected Corridors*, Washington, DC: World Bank.
Raballand, Gaël (2003), 'Determinants of the Negative Impact of Being Landlocked on Trade: An Empirical Investigation Through the Central Asian Case', *Comparative Economic Studies*, 45, 4, pp. 520–536.
Stark, O. (1991), *Economics of Migration*. Boston: MIT Press.
Todaro, M. (1976). *Internal Migration in Developing Countries*. Geneva: ILO.
UN (2005), 'Landlocked Developing Countries', online at: http://www.un.org/special-rep/ohrlls/lldc/default.htm, accessed at 1 April 2012.

World Bank (2005), 'Workers' Remittances to Developing Countries: A Survey with Central Banks on Selected Public Policy Issues', *World Bank Policy Research Working Paper* No. 3638.

World Bank (2006), *Global Economic Prospects 2006: Economic Implications of Remittances and Migration*, Washington, DC: World Bank.

World Bank (2006a), *World Development Indicators*, Washington, D.C.: World Bank.

World Bank (2007a), *Migration and Remittances in the Former Soviet Union* (eds.) Bryce Quillin Carlo Segni Sophie Sirtaine Ilias Skamnelos. Washington, DC: World Bank.

Zaionchkovskaya, Z.A. (2007), 'Migration Patterns in the Former Soviet Union', online at: http://www.rand.org/pubs/conf_proceedings/CF130/CF130ch2.pdf.

Are Hofstede's National Cultures Homogeneous? The Case of Uzbekistan

Munira Aminova and Marc Jegers

1. INTRODUCTION

The culture of any country is unique so the method of measuring or quantifying culture with its numerous artefacts, values and norms is pivotal. There are wide-ranging and contradictory scholarly opinions about what should be measured to represent the concept of *culture*. A common fashion is to measure culture with culture indexes developed by Hofstede (1976). The typology of Hofstede has created "a paradigm shift á la Thomas Kuhn" (as quoted in Hoppe 2004), it is regarded by scholars as path-breaking, astounding, classic, and a standard against which the new work on cultural differences is validated (Ailon 2008). However, the validity of this approach has been extensively debated (McSweeney 2002; Barkema and Vermeulen 1997; Hunt 1981). Territorial uniqueness of culture is another problem identified by them. Some scholars like Straub *et al.* (2002) Explore a theory-based view of culture via social identity theory (SIT), which suggests that each individual is influenced by a plethora of cultures and sub-cultures, some ethnic, some national, and some organizational.

Others have tried to use broader cultural categories such as *Asian* (Baughn *et al.* 2006). *European* and others; though normally this represents an amalgamation of different countries rather than a cross-cutting category. Despite the large number of criticisms it is still accepted that Hofstede was able to measure, quantify and graph the so called culture into four dimensions: power distance, individualism, masculinity/feminity, and uncertainty avoidance. During the fourteen years of use three more were added: long-term orientation, indulgence/restraint, monumentalism. Hofstede himself admitted that states are not the best units for studying cultures, but "they are usually the only kind of units available for comparison and better than nothing" (Hofstede 2002).

The analysis of culture in Uzbekistan is a case in point. Firstly, it is not clear which cultural grouping to fit Uzbekistan into – how *Asian* is a country that for about a century was ruled from Moscow, only becoming independent in 1991. Likewise, while the country's population is predominantly Muslim, again the experience of being largely isolated from the larger Muslim world when part of the Soviet Union suggests that the impact of religious affiliation differs from that in other parts of the Muslim world (Stevens 2006).

This paper analyses the cultural clusters of Hofstede by applying them into two different categories of people residing in Uzbekistan. The paper introduces the context of Uzbekistan with some brief description of culture, traditions and religion. It continues by describing the methodology and eventually provides results and analysis.

2. UZBEKISTAN

Since ancient times, the wealth of Central Asia has attracted foreign conquerors such as Alexander the Great (356–323 B.C.), Gengizkhan (1206–1227) and many others. Whether Greek, Persian, Turk, or Mongol, the worriers left the remnants of their cultures. In the mid of 13th century local dynasties came to power. During that time trade flourished and the market was dominant. Uzbekistan has inherited an exceptionally rich cultural legacy, both tangible and intangible, that dates back thousands of years (Kurbanova 2006). It is a blend of traditions, religion, rituals and cultural artefacts. The culture has been also shaped due to the country's privileged position at the crossroads of the ancient Silk Road; cultural traditions have been enriched by the interactions between Eastern and Western civilizations.

"The country is a syncretic one, with Buddhist stupas, buried in the sand, dotting a landscape of mosques that repeat Buddhist forms; Islamic tombs decorated with Pythagorean geometrical designs; and the Persian new year of Navruz a state holiday. Uzbeks drink from Chinese-style teacups, wear Turkish-style skullcaps, and adhere to Arab religious traditions. Their culture also reflects an Eastern mysticism of dervishes and shrines, as well as centuries of Jewish settlement and the influx of traders from across the Eurasian continent" (Jarvik 2006)

Uzbekistan with a population of 29 million (2012 estimates) comprises nearly half (roughly 45%) of Central Asia's total population. Present-day Uzbekistan is aspiring to build a national multi-ethnic state or 'ethno-centric state' (Koroteyeva *et al.* 2004), based on geography rather than language, religion, or race; and a culture that is used to describe a national way of life. This is done by promoting local *cadres*, introducing national holidays, changing priorities in history and other realms. The policy of the government is to 'revive national tradition' rather than to build a new state.

Uzbekistan consists of twelve administrative units, called *viloyats*. All of them have considerable differences in their Uzbek dialect. There are few major cities like Samarqand and Bukhara where the spoken language is Tajik.[37] This

---

37 While having discussions on language it is important to note the reforms in script which have taken place within one century in Uzbekistan. Before the Russian invasion the local

originates from the Persian language group, whereas Uzbek belongs to the Turkish languages group. There are a number of smaller cities within Uzbek speaking viloyats like Nurata within Navoi Viloyat or like Chust within Namangan viloyat and others that also speak Tajik.[38]

There are also wide differences in traditions like marriages, which in Uzbekistan is a set of complex rules and procedures that have to be followed.[39] For instance in Khorezm Viloyat a groom has to pay the father of a bride to marry her[40], whereas in Tashkent city a wedding is quite costly for the family of the bride. The preparation of a national dish called *plov* is also quite different in viloyats (Aminova 2011).

Religion: Uzbekistan has a predominantly Muslim population, during soviet times religion was discouraged or even persecuted. So people stopped practicing it widely, but they still kept some essential 'religious rituals' rather referred to as 'community practices' that they performed during weddings, deaths, child births and others. After independence religion had a potential to fill the vacuum in the absence of a national ideology and because of the large disorientation of the population. However, after series of bombings in 1999 religion was considered a threat to stability, and strict rules were introduced on anyone heavily involved in religion. Religious tolerance is different in the twelve viloyats. In Samarkand for instance tolerance is higher for alcohol, pork and relaxed approach to clothing style; in Ferghana region the rules are much stricter. Uzbeks consider themselves Muslims and are proud to be. At the same time local community is quite tolerant for other religions; there are Christians, Buddhists and others, who have their churches around the country.

---

script was Arabic. In the 1920s, it was decided that all languages written by Soviet Muslims should shift to Latin script. In the 1930s, it was decided to shift from Latin to Cyrillic script. After the independence in the 1991s Latin script was introduced again. Older generations are illiterate in Latin. Repeated script reforms have complicated language and literacy development in Central Asia according to researchers like Bahry *et al.* (2006), Schlyter (2004), Landau and Kellner-Heinkele (2001) and Fierman (1991).

38 There is a certain dislike among the two language carriers but so far it did not cause any major conflicts within Uzbekistan. According to some sources about 20% of population consists of Tajiks. After the dynasty of Somonids in the region the official government language was Farsi, which is known as Tajik nowadays. Culture, arts, poetry and governance were in Farsi. According to Koroteyeva and Makarova (2005) Tajiks do not deny that they are grown up in a Tajik environment and speak Tajik language but consider themselves citizens of Uzbekistan and refer to themselves as Uzbeks sometimes.

39 Arranged marriages are still a norm in Uzbekistan and mostly seem to comfort everyone involved.

40 Normally these days it is a nominal price, in order to follow the tradition.

3. METHODOLOGY

This paper uses the variables derived from cultural clusters of Hofstede's Values Survey Module (VSM) Questionnaire© (2008) designed on the basis of a survey of IBM workers conducted in 1968–1973 which became the de-facto standard in cross cultural research. It was conducted in Uzbekistan in 2008.

Two different groups of people participated in this research: The first group consisted of people who had considerable international exposure (living or studying abroad or who have just returned) but who were born and had grown up in Uzbekistan, and consider themselves Uzbeks (the questionnaire was filled both online and on paper). The second group consisted of people who lived and worked in Uzbekistan only and work in standard semi-government/government organizations (the questionnaire was translated into Russian).

Uzbekistan was chosen because it is quite a unique case; it is a Muslim country, ruled by Moscow for some 70 years. It is also a post-soviet transitional country with quite a unique culture which is considered as a unique hybrid of national and Muslim symbolism nurtured throughout centuries. Personal networks of one of the authors enabled easy access to primary data.

3.1 Hofstede's dimensions

3.1.1 Power distance index (PDI)

PDI refers to national values with respect to power inequality in the workplace and in the society at large. "It reflects the degree to which an unequal distribution of power and authority in institutions is viewed as legitimate" (Ringov and Zollo 2007).[41] PDI is the extent to which the less powerful members of organizations and institutions (like the family) accept and expect that power is distributed unequally (Hofstede 1980). There is evidence that people from high power distance cultures would consider a questionable business practice more ethical than people form low power distance cultures (Cohen *et al.*, 1996). The formula used for scores on VSM for identifying a score for Power Distance:

$$PDI = 35(m07 - m02) + 25(m23 - m26) + C(pd)$$

C=50 in all formulas

---

41 With respect to the formulas mentioned in this chapter, see the explanation in the Annex.

3.1.2 Individualism (IDV)

The IDV dimension relates to the societies where ties between individuals in the societies are loose: everyone is expected to look after him/herself and his/her family. On the collectivist side, we find societies in which people from birth onwards are integrated into strong, cohesive in-groups, often extended families (with uncles, aunts and grandparents) which continue protecting them in exchange for unquestioning loyalty (Hofstede 1980). The formula used for scores on VSM for identifying a score for Individualism:

$$IDV = 35(m04 - m01) + 35(m09 - m06) + C(ic)$$

3.1.3 Masculinity (MAS)

Highly masculine societies place low value on caring for others, on inclusion cooperation, and solidarity. Career advancement, material success and competition are paramount. Cooperation is considered as a sign of weakness (Ringov and Zollo, 2007). The formula used for scores on VSM for identifying a score for Masculinity:

$$MAS = 35(m05 - m03) + 35(m08 - m10) + C(mf)$$

3.1.4 Uncertainty avoidance (UAI)

UAI deals with a society's tolerance for uncertainty and ambiguity. It indicates to what extent a culture´ programs its members to feel either uncomfortable or comfortable in unstructured situations. Unstructured situations are novel, unknown, surprising, and different from usual. Uncertainty avoiding cultures try to minimize the possibility of such situations by strict laws and rules, safety and security measures (Hofstede, 1980). The formula used for scores on VSM for identifying a score for Uncertainty avoidance:

$$UAI = 40(m20 - m16) + 25(m24 - m27) + C(ua)$$

3.1.5 Long term orientation (LTO)

This dimension appeared in the second round of the Hofstede's research and was designed by Chinese scholars. The values here are thrift and perseverance for LTO and respect for tradition, fulfilling social obligations, and protecting one's 'face' for the Short Term Orientation. The formula used for scores on VSM for identifying a score for LTO:

$$LTO = 40(m18 - m15) + 25(m28 - m25) + C(ls)$$

The other two dimensions, which we will not stress but still provide a result, are *Indulgence versus Restraint Index and Monumentalism* Index (MON) which were included in VSM 2008 based on the work of Minkov (2007). The formulas used for scores on VSM for identifying a score for IVR and MON are:

$$IVR = 35(m12 - m11) + 40(m19 - m17) + C(ir)$$

$$MON = 35(m14 - m13) + 25(m22 - m21) + C(mo)$$

4. RESULTS

4.1. General characteristics

The gender distribution in both samples is quite similar. There are 68% and 67% male respondents; the ratio of men/women is 1000/1007 (UzStat 2001) in Uzbekistan. However, the majority of the employed in the institutions were men. The age range of respondents was generally between 20–40 years. In Uzbekistan 43% of the population is under the age of 20 and only 5.8% is over the age of retirement (UzStat 2001).

Both samples include people with a higher education being pursued or finished. Especially in Sample 2 we can see that the majority of respondents have extensive experience in educational establishments.

Table 1: Occupation of respondents

	Sample 1 (%)	Sample 2 (%)
No paid job (includes full-time students)	4	4
Unskilled or semi-skilled manual worker	0	2
Generally trained office worker or secretary	0	10
Vocationally trained craftsperson, technician, IT-specialist, nurse, artist or equivalent	8	18
Academically trained professional or equivalent (but not a manager of people)	64	30
Manager of one or more subordinates (non-managers)	20	20
Manager of one or more managers	4	16

There are about 4% of student respondents in each sample; largely most of the respondents have professional or administrative jobs.

4.2. Nationalities of respondents

Uzbekistan used to be a very multinational country. Even now there are a lot of people of various nationalities, who were born and raised in Uzbekistan, Uzbek 80%, Russian 5.5%, Tajik 5%, Kazakh 3%, Karakalpak 2.5%, Tatar 1.5%, other 2.5%.[42] The country has an ethnic Korean population that was forcibly relocated to the region from the Soviet Far East in 1937-1938 (UzStat 2008). All of them reside in the territory of Uzbekistan. It is difficult avoiding the presence of different nationalities in the survey. However, all of the nationalities are blended into society so well that we almost do not see the difference in daily life.

Table 2: Nationalities of respondents

Nationalities of respondents	Sample 1 (%)	Sample 2 (%)	Country average
Uzbeks	91.0	65.0	80
Russian	9.0	17.0	5.5
Korean		9.0	
Tatar		3.0	1.5
Kazakh		4.0	3
Jewish[43]		2.0	
Tajik		0	5

The differences in nationalities also might have affected the answers, although the respondents were chosen between those who were born and grown up in Uzbekistan. Especially respondents of Sample 2 are those who rarely if at all travel abroad.

The two seemingly not so different and culturally similar groups produced strong differences in the results regarding Power distance, Individualism and Uncertainty avoidance. To check the significance of the differences between each sample's values for the five cultural dimensions, t-values for the observed differences in means were estimated assuming equal variances (per dimension)

---

42 There were 94,900 Jews in Uzbekistan in 1989; about 0.5% of the population according to the 1989 census), but now, since the collapse of the USSR, most Central Asian Jews left the region for the United States or Israel. Fewer than 5,000 Jews remain in Uzbekistan.

in both samples, as we could only determine the variance in one sample, due to the unavailability of the raw data of Sample 1. Item averages were substituted for missing values. The t-values obtained for all except one cultural dimension (LTO) were so high ($p < 0.001$) that we can safely conclude that genuine cultural differences exist between the two sample populations.

Table 3 presents the outcomes of the indulgence/restraint (IVR) and monumentality (MON) scores, for documentary purposes.

Table 3: IVR and MON Indexes

	Sample 1	Sample 2
Indulgence versus Restraint	86.08	63.60
Monumentalism	90.11	97.39

Every member of a nation exhibits national characteristics, so it matters little which individual we study (Farber, 1950 quoted in McSweeney, 2002). However, from the above results we can see that there are significant intra-country differences between the two samples which indicate that national culture is not identically common to all national individuals. Of course we can combine the two samples and have an average value; in heterogeneous sets of data it is always possible to find a central tendency and admit the average is what we mean by a culture of that particular country. However, as Schwartz et al. (1992) argue, we cannot derive normative ideas of a culture from the average of individual responses.

It is widely accepted that a culture change does not happen overnight, but the extensive changes during the rapid transformation from socialism to a market economy had substantial impact on society. The entire generation of people brought up under socialism became 'handicapped' and could not adapt to new realities. Some who were able to adapt quickly are well off, most others are underemployed; their skills, knowledge and value system became obsolete.

4.3. Power distance

The PDI equals to 61.22 in a Sample 1, and 22.14 in Sample 2. Despite the common belief that Uzbekistan is a country where both subordinates and leaders accept the power distance. There is a certain element in the culture called 'hurmat' deference, which means everyone who is older than you should be respected. For example, youngsters should always greet elders first[44] even if they do not know them personally; never express non agreement by saying directly 'no' to an elder (this includes people holding higher positions too). However, our results contradict this myth. Power distance in Uzbekistan also depends on personal and social networks. If one is a relative or a friend of a 'well-respected' person that knows your direct boss, the power distance could be lowered.

The results also show that power distance is related to the size of the organization. The larger the organization, the more important is the power distance and vice versa. Hierarchical structures of a smaller organization allow more possibilities for a superior to consult his employees, whereas in large organizations this is rather difficult. Since many organizations in Uzbekistan still use the organizational structures inherited from Soviet times, 'dual subordination' is not abnormal. Even if the questionnaire asks them to think about their ideal jobs, the present job still affects the answers.

Power Distance relates to the notion of leadership; how each culture would regard leaders. In France for instance the CEO of the company should be referred to as Monsieur President until his death. In Uzbekistan it is difficult to imagine that the 'big boss' would have lunch or dinner with subordinates/employees, or would socialize with them. CEOs in most cases will go out with the people of the same level or of the same social network.

4.4. Individualism

Results of the survey and questionnaire brought unexpected results for the individualism index too. Being a part of a former communist country and partly influenced by the collectivistic nature of the Islam religion, it was expected that the Individualism Index would be much lower in both samples. However, what we found is 73.29 for Sample 1 and as high as 99.82 for Sample 2.[45]

If we look at Uzbekistan independently from the questionnaire, it presents the image of a very collectivistic society; people like to be and spend time in groups. For instance, someone eating out alone is considered as strange behavior. People have large families and belong to family clans all their lives.

---

44 We refer to elders in this part of the paper as to anyone who is considerably older, for instance a teenager will be older for children, or adults/parents for school children, or old people/pensioners for everyone.
45 Technically, it is difficult to say whether or not such values are high, since we cannot compare them with other countries. We can only see the difference between samples and other dimensions.

Extended family members like uncles, aunts and their children, grandparents and their relatives are all considered as a part of the family. Weddings and family events, therefore, are quite large and expensive. The relationships are based on loyalty and trust.

In addition, anyone who is born in Uzbekistan automatically belongs to a so called *Mahalla*. This is an Arabic word meaning *local community* and refers to a community of people residing in a specific territory. Mahallas may vary in size from 150 to 1,500 families (OSI 2000). Some researchers say it is "the neo-traditional community of Uzbekistan" (Masaru 2004), others call it a "national school of Eastern democracy" (Karimov 1994). Most important events in the life of a three-generation Uzbek family (three generation families are standard for Uzbeks) occur with the assistance and direct participation of mahalla (Masaru 2004).

However, the drawbacks are that group decision making is not valued. And creating group atmosphere and team spirit is not a strength of Uzbek culture. It rather concentrates on the integration of family members into business and creating a trustful environment especially when the formal rules and procedures and the rule of law are quite weak (Aminova 2011). Uzbek people are very (extended) family oriented. However, we cannot see any form of workers union for instance that is created in the country that people could be loyal to.

4.5. Masculinity

In Uzbekistan the focus is more on assertiveness, competitiveness, wellbeing of the person and his family members rather than the quality of life and social welfare as in Nordic countries like Sweden and Denmark.

The importance of task versus relationship can also be considered while analyzing how decisions are made in Uzbekistan. Like in Asian, Latin American and Middle Eastern countries, in Uzbekistan a relationship must be established before the business can be conducted. Companies prefer to hire a relative rather than a non-related person. Without a foundation of trust it appears to be impossible to engage a person.

Masculinity is demonstrated in having big houses and big cars and organizing posh and expensive ceremonies for marriages or funerals, birthdays and other occasions. The continuing influence of the Soviet culture on forms of observance and social attitudes, and the conviction that women have a role to play in the public realm formed a class of educated elites in Uzbekistan, with a high education level and medical care network and specialists. However, despite the formal acceptance of the role of women in society there is an invisible separation between the two genders that effectively prevents women from taking an active role in public governance or in business. In a society which largely depends on extended personal and social networks in conducting a business or getting promotion in the public sector or any large organization per se, and where notions of 'face saving' and shame are used for social control the inclusion of women into governance and decision making requires a major shift in thinking or attitude. Culturally minors and women are protected and taken care of by 'stronger' – men. It is possibly a sense of social responsibility that motivates people to take care of each other. If a neighbor is sick it is very normal for example to cook food and share it with her/him. If there is a funeral, the grieving family is not allowed to cook for up to 3 days. It would be relatives or neighbors who will bring food to them during that time.

On the other hand, charitable giving can also be traced back to Islamic religion. It survived the Soviet years by being integrated in the mahallas. Other researchers argue that it is a community custom (in Uzbekistan) that those who are better off share their wealth with their neighbors and citizens accordingly provide material assistance to the community. Traditionally, these acts are performed on a voluntary basis and are not widely publicized.

4.6. Uncertainty avoidance

According to Hofstede (1983) we can create security in three ways: technology (used to protect ourselves against nature and war); laws (we design laws and spell out procedures for all possible situations), and religion (all religions make uncertainty more tolerable).

Uzbekistan is a low uncertainty avoidance country. Strict procedures and rules were inherited from Soviet times, however, at present day rules and regulations bend easily; it is rather the 'rule of man' rather than a 'rule of law' guides the procedures of the country and individual organizations. The will of the father in the family or the manager in the organization is more important than the written rules. Uzbek people like to build big houses but that is rather a sign of power then a self-protecting measure.

5. CONCLUSION

This research suggests that there may be very large intra-country differences in culture and that the country boundaries are not always good demarcation lines, especially if the borders have been somewhat artificially created by colonization, conquest or other. Analyzing the culture of Uzbekistan seems a case in point, because it had such a wide influence of other cultures during the course of time.

REFERENCES

Ailon, G. (2008), 'Mirror, mirror on the wall: culture's consequences in a value test of its own design,' *The Academy of Management Review, 33*(4), pp. 885–904

Alasuutari, P. (2001), 'Art, Entertainment, Culture, and Nation,' *Cultural Studies Critical Methodologies, 1*(2), pp. 157–184.

Aminova, M. (2011), 'Informal structures and governance processes in transition economies: the case of Uzbekistan,' *International Journal of Public Administration,* 34, 9.

Bahry, S., Niyozov, S., and Shamatov, D. (2008), 'Bilingual Education in Central Asia,' in N.H. Hornberger (eds.), *Encyclopedia of Language and Education,* Berlin: Springer, pp. 1655–1671.

Barkema, H.G. and Vermeulen, F. (1997), 'What Differences in the Cultural Backgrounds of Partners are Detrimental for International Joint Ventures?' *Journal of International Business Studies, 28* (4), pp. 845–864.

Baughn, C.C., Nancy L.B., and McIntosh J.C. (2007), 'Corporate Social and Environmental Responsibility in Asian Countries and other Geographical Regions,' *Corporate Social Responsibility and Environmental Management 14*(4), pp. 189–205.

Cohen, J.R., Pant L.W., and Sharp, D.J. (1996), 'A Methodological Note on Cross-Cultural Accounting Ethics Research,' *International Journal of Accounting 31*(1), pp. 55–66.

Hofstede, G. (1976), 'Nationality and espoused values of managers,' *Journal of Applied Psychology, 61*(2), pp. 148–155.

Hofstede, G. (1980), *Culture's consequences: International differences in work-related values.* Thousand Oaks: Sage Publications.

Hofstede, G. (2001), *Culture's consequences: comparing values, behaviours, institutions, and organizations across cultures.* Thousand Oaks: Sage Publications.

Hofstede, G. (2002), 'The pitfalls of cross-national survey research: A reply to the article by Spector *et al.* on the psychometric properties of the Hofstede Values Survey Model 1994,' *Applied Psychology: An International Review, 151*(1).

Hoppe, M.H. (2004), 'An interview with Geert Hofstede,' *Academy of Management Executive, 18*(1), pp. 75–79

House, J.R., Hanges J.P., Masour J., Dorfman P. and Vipin G. (2004), *Culture leadership and organizations: The Globe study of 62 societies.* Thousand Oaks: SAGE Publications.

Hunt, J.W. (1981), 'Applying American Behavioural Sciences: Some Cross-Cultural Problems,' *Organizational Dynamics, 10*(1), pp. 55–62.

Jarvik, L.A. (2005), 'Uzbekistan: A Modernizing Society,' *Orbis: A Journal of World Affairs 49,* pp. 271–273.

Karimov A.I. (1994), *Ideologiya nacionalnoy nezavisimosti – ubejdeniya naroda i vera v velikoye budushee,* Tashkent: Uzbekistan Press.

Koroteyeva, V., and Makarova, E. (1995), *The Assertion of the Uzbek National Identity,* International Institute of Asian Studies Publications.

Kurbanova, D. (2006), 'Rich Traditions Cement Family and Community Ties in Uzbekistan Time-honored ceremonies continue to inspire Uzbeks,' *SangSeng,* 46–47.

Masaru, S. (2004), 'The Politics of Civil Society: Mahallyas and NGOs of Uzbekistan,' *Slavic Research Center 10*, 335–368.

McSweeney, B. (2002), 'Hofstede's Model of National Cultural Differences and Their Consequences: A triumph of faith - A failure of analysis,' *Human Relations* 55(1), pp. 89–118.

Myers, D.G. and Diener E. (2006). 'Who is Happy?' *Psychological Sciences,* 6(1), pp. 10–19.

Ringov, D. and Zollo, M. (2007), 'The impact of national culture on corporate social performance,' *Corporate Governance,* 7(4), pp. 476–485.

Schneider C. S. and Barsoux J.L. (2003), *Managing Across Cultures,* London: Prentice Hall.

Spechler, M. (2008). *The Political Economy of Reform in Central Asia: Uzbekistan under Authoritarianism.* London and New York: Routledge.

Spyridakis, J.H., Wei, C., and Kolko, B.E. (2003), 'The relationship of culture and information-seeking behaviour: A case study in Central Asia,' in *Adjunct Proceedings of HCI International 2003,* Crete, Greece: Crete University Press, pp. 167–168.

Straub, D., Loch, K., Evaristo, R., Karahanna, E., and Srite, M., (2002), 'Toward a Theory-Based Measurement of Culture,' *Journal of Global Information Management,* 10(1), pp. 13–23.

Stevens, D. and Mukhamedova, L. (2008), 'Social Responsibility at the grassroots – the influence of '*mahalla*' community organisations on the CSR practices of small and medium enterprises (SMEs) in Uzbekistan,' in A. Aras, G., and Crowther D. (Eds.), *Culture and Corporate Governance.* Social Responsibility Research Network.

Annex: General characteristics of the respondents

	Sample 1 (%)	Sample 2 (%)	2006 Uzbekistan
Gender of respondents			
Male	68	67	49.9
Female	32	33	50.7
Age of respondents			2007
Under 20	2.0	3.0	43.6
20-24	36.0	29.0	10.5
25-29	28.0	29.0	8.4
30-34	16.0	23.0	7.4
35-39	10.0	11.0	6.4
40-49	6.0	3.0	11.5
50-59	2.0	0	6.3
60 or over	0	0	5.8
Years of education of respondents			
10 years or less	2	0	
11 years	8	0	
12 years	8	0	
13 years	6	5	
14 years	12	14	
15 years	22	20	
16 years	12	14	
17 years	20	11	
18 years or over	10	32	

Sources: The State Statistics Department of the Ministry of Macroeconomics and Statistics of the Republic of Uzbekistan, 'Women and Men of Uzbekistan: Statistical Collection,' 2002;
online: http://www.statistics.uz/data_finder/2344/pdf/population_age-based.pdf.

Institutional Change and Agricultural Performance in Kyrgyzstan

Kamiljon T. Akramov and Nurbek Omuraliev

1. INTRODUCTION

There is a consensus among economists that the transition from a command economy to a market economy is a large-scale institutional change (Murrell 2008). In the agricultural sector, this transition involves several distinct aspects of institutional change, including the decollectivization and individualization of land use, the introduction of private property rights, and the building of market and collective action institutions (Lerman *et al.* 2004; Roselle and Swinnen 2004; Spoor 2003). In most transition countries, including Kyrgyzstan, agricultural development has been determined by the success of these changes. Nevertheless, there is ongoing debate concerning the effects of decollectivization and individualization on agricultural performance. On the one hand, the literature claims that individualization of agriculture in transition economies increased productivity by solving the incentive and governance problems pertinent to large farms. Evidence suggests that transition countries with higher shares of land in individual use have achieved better results in agricultural growth and productivity (Lerman *et al.* 2004; Lerman 2004, 2008; Macours and Swinnen 2002). On the other hand, some argued that individualization of agriculture may lead to subsistence farming, which is considered to be a survival strategy and is usually associated with low productivity (Sarris *et al.* 1999). Policymakers in some transition countries blame the individualization of agriculture for the fragmentation of farmland and disorganization in the supply chain.[46] The literature also suggests that individual farmers, in addition to access to land, need access to working capital, input and output markets, and traditional and new agricultural services (Deininger 2002; Spoor 2003). In this regard, Kyrgyzstan's experience is of interest, where despite the far-reaching land reform implemented in the mid-1990s individual farmers still experience major problems related to market and collective action institutions.

Agriculture is an important sector of Kyrgyzstan's economy, employing more than 40% of the country's labor force and generating about one-third of its GDP. Its transformation began in the early 1990s and initially focused on the liberalization of agricultural markets and prices. Starting in the mid-1990s, the

---

46 This assertion is based on the authors' personal observations.

agricultural reform concentrated on land reform and the individualization of agricultural production. Major changes during this period included the abolition of state and collective farms, and the transition to a private farming system through equitable land distribution and the introduction of private ownership of agricultural land. Another important change was the establishment of institutions for collective action, such as water users associations that operate and maintain on-farm irrigation facilities and regulate water allocation.

In the early 1990s, Kyrgyzstan experienced a dramatic decline in agricultural output. However, agricultural production rebounded quickly following the land reform, and showed a clear upward trend between 1995 and 2001. Indeed, during this period, agricultural performance in Kyrgyzstan (in terms of gross output, increased labor productivity and improved yields) was one of the best among all former Soviet republics (Roselle and Swinnen 2004). The rates of agricultural growth, however, slowed significantly after 2002. There is widespread concern among policymakers in Kyrgyzstan that significant declines in agricultural growth rates are associated with the fragmentation of land use and the apparent inability of small peasant farms to sustain growth in agricultural productivity. They argue that "small peasant and individual farms are not able to apply high technology of production because of the small sizes of the farms, low income and lack of access to high quality agricultural machinery" (Government of Kyrgyzstan, 2004, p. 10). In fact, the evidence presented in this paper shows that small peasant farms face difficulties in accessing agricultural machinery and inputs. Another essential problem in Kyrgyzstan's rural sector is that the decollectivization of agriculture was accompanied by a significant decline in agricultural public expenditures, leading to significant declines in the availability of agricultural services.

In this paper, we first review the major agricultural reforms that significantly changed the institutional structure in rural Kyrgyzstan and examine its impact on agricultural performance. Then we discuss the remaining challenges and constraints in the agricultural sector that prevent Kyrgyz famers from fully utilizing the benefits of private land ownership. The impacts of institutional changes on agricultural performance are examined using official statistical data from both national and international sources. The constraints to agricultural growth are analysed using secondary data from a recent farm survey.

2. INSTITUTIONAL CHANGE IN RURAL KYRGYZSTAN

Prior to 1991, land in Kyrgyzstan as elsewhere in the former Soviet Union was solely state owned and the agricultural sector was dominated by large-scale socialist farms. As of January 1, 1991, there were several hundred large-scale state and collective farms in Kyrgyzstan. These farms managed over 15 million hectares of agricultural land, including about 1.3 million hectares of arable land, perennials and pastures (Bloch *et al.* 1996). Alongside these large farms, hundreds of thousands of rural households cultivated small plots, collectively accounting for about 3-4% of the country's arable land. Despite their small share in agricultural land, household plots achieved relatively high levels of productivity. According to official statistical data from the National Statistics Committee of Kyrgyzstan (NSC), in 1990, the household sector produced about two-thirds of the country's total production of fruits, over 50% of potatoes and milk, over 45% of meat, and about 40% of vegetables. State and collective farms, on the other hand, dominated the production of cereals and technical crops (NSC 2012). The transformation of the agricultural sector in Kyrgyzstan began in the early 1990s and continues to the present day. This transformation has included several distinct aspects of rural institutional change, including the abolition of large-scale farms, the distribution of agricultural land to individual farmers and introduction of private property rights to agricultural land, and the creation of market and collective action institutions.

2.1 Farm Reorganization and Land Reform

The collapse of the socialist economic system and the shift to a more market-oriented system in the early 1990s created the need for structural and institutional reforms in the country. In the earlier phase of rural reform (1991–1995), the government eliminated most of the state subsidies for agricultural inputs, and deregulated the markets and prices for agricultural output. However, initial farm reorganization was limited to the restructuring of state and collective farms into agricultural cooperatives and peasant farm associations (Bloch *et al.* 1996). While these reform attempts started quickly, they were inconsistent. An examination of the farm structure that existed in 1995 provides evidence of the limited progress made during the first half of the 1990s. After four years of reform, only 12% of Kyrgyzstan's arable land was under cultivation by individual farms, while the rest of the arable land was still controlled by large agricultural enterprises (NSC 2012). These inconsistent reforms and significant declines in terms of trade in the agricultural sector (due to large differences in agricultural input and output prices) led to a dramatic drop in agricultural production. As a result, poverty dramatically increased in the country during this period (Anderson and Pomfret 2003).

The second phase of the reform, which started in the mid-1990s, focused on land reform. In the aftermath of the significant declines in agricultural production, the Kyrgyz government began to implement historic institutional changes in the form of land distribution. Starting in late 1994 and through most of 1995, the government passed several legal and policy directives[47] that introduced serious measures aimed at dismantling most of state and collective farms. Within a short period of time, more than 450 state collective farms were liquidated and 75% of all agricultural land (except pastures) was distributed to eligible rural people[48] under an equity principle (USAID 2008). The amount of land allocated to each individual depended upon the number of eligible people living in the vicinity of the state or collective farm, the size of the farmland, and the years of experience of the farm workers. The resulting land holdings varied from 0.1 ha/person to 1 ha/person, with the smallest holdings located in the more densely populated southern provinces (Gioveralli 1998).

About 25% of the agricultural land was preserved in state ownership as part of the Agricultural Land Redistribution Fund (LRF) under the Ministry of Agriculture and Water Resources (MAWR) of Kyrgyzstan. The management of the LRF land was given to local councils[49], who were tasked with leasing this land to individual farmers through auction, tender or direct allocation. This process is now regulated by the Model Regulation on the Conditions and the Procedure of Leasing out of LRF Land[50] adopted by the Parliament (USAID 2008).

Initially, farmers received agricultural land use certificates of up to 99 years.[51] These land use certificates granted five legal rights to individual farmers, namely the rights to transfer, exchange, sell, lease, and use the land as collateral for credit. In 1998 a country-wide referendum adopted a constitutional amendment that allowed private land ownership. As a result, all land-use certificates were converted into private land ownership documents. However, an

---

47 The most important governmental decrees on rural reform passed during this period include: the 1994 Presidential Decree on Measures for Intensification of Land and Agrarian Reform; the Regulation on the Procedure for the Determination of Land, and the Regulation on Reorganization of Agricultural Enterprises.
48 Individuals eligible to receive land shares included those working and living on a given farm, those retired from or disabled by work on a given farm, persons born on a given farm working elsewhere who decided to return and take up permanent on-farm residence, and people living on the farm but working in other sectors.
49 In 1996, these local councils were transformed to local self-government units (called *Ayil Okmotu*) as part of the comprehensive decentralization reform initiative.
50 The initial Model Regulation was adopted by Parliament on April 15, 2002. The revised version of the Model Regulation was developed with assistance from USAID and adopted by Parliament on June 29, 2007. The regulation requires all LSGs to develop a Strategic Plan (including classifications and a map of the LRF land) for the LRF's use (USAID 2008).
51 Local self-governments in rural areas were responsible for issuing formal certificates of land use rights.

important amendment to the Land Code (in 1999) introduced a five-year moratorium for sales and purchases of agricultural land. This moratorium was lifted as of September 1, 2001 with the adoption of a law entitled "On Agricultural Land Regulation." Currently, agricultural land may be owned by the state, by citizens of Kyrgyzstan who are at least 18 years old and have permanently resided in rural areas for at least two years, and by agricultural cooperatives. The agricultural land shares and parcels may be leased out, sold or donated, but must be used exclusively for the purposes of agricultural production. The land shares and parcels may be traded for other land shares or parcels, but only within the boundaries of a given local self-government (LSG).

2.2 Irrigation Reform and Water Users Associations

Irrigated agriculture, which comprises more than 1 million hectares of arable land in Kyrgyzstan, is a major component of the country's rural economy. However, the irrigation system in Kyrgyzstan was originally built to serve large-scale farms and was not suitable for the needs of numerous small-scale farmers that emerged following the land reform. Prior to reform, the irrigation system was managed jointly by the Rayon (district) Irrigation Departments (RIDs) and the former state and collective farms. All inter-farm and off-farm irrigation networks, including the main canals, were operated and maintained by the RIDs and funded through the state budget. The RIDs were responsible for delivering irrigation water through the off-farm and inter-farm networks to the head gates of state or collective farms. In contrast, the on-farm irrigation systems were operated and maintained by state and collective farms (Johnson and Stoutjesdijk 2008).

The individualization of agricultural land use in the mid-1990s created an institutional vacuum, as no organization was responsible for the operation and maintenance of the on-farm irrigation networks. This problem was heightened by the collapse of public funding for the operation and maintenance of such networks, and the subsequent deterioration of the inter-farm irrigation systems. In order to address this problem, the Kyrgyz government introduced two important measures. First, in August 1994, the government gave the local councils the ownership of and the responsibility for operating and maintaining on-farm irrigation infrastructures. It was assumed that the local councils would collect land taxes from farmers and use a portion of these revenues to maintain the on-farm irrigation infrastructures. Second, in 1995, the government instituted an irrigation service fee that was to be paid to irrigation water providers, and used to operate and maintain the off-farm irrigation networks.

However, merely introducing these measures proved to be insufficient to ensure the sustainable management of irrigation infrastructure. The local councils did not have adequate staff or funds to fill the institutional vacuum, making it unfeasible for them to properly operate and maintain the on-farm irrigation networks. As a result, the on-farm irrigation infrastructures continued to deteriorate. Lately, donors and governments have widely supported collective action management solutions based on water users associations (WUAs) (Johnson and Stoutjesdijk 2008). With the help from international development institutions, the Kyrgyz government enacted a resolution on the establishment of WUAs on August 13, 1997. This resolution was driven by the state's motivation to encourage collective action and decentralize the formerly centrally managed irrigation sector by involving farmers in the management of on-farm irrigation networks. However, this resolution was not sufficient to provide a legal framework for creating WUAs and transferring on-farm irrigation infrastructure to the ownership of formally established WUAs. Later, the law entitled "On Unions (Associations) of Water Users," which was adopted in early 2002, reflected the legal status and organizational basis for the establishment of WUAs as non-commercial organizations to operate and maintain irrigation infrastructure in rural areas.

The new Water Code of Kyrgyzstan, which was adopted on January 12, 2005, provided a legal basis for further giving farmers the responsibility for operating and maintaining the irrigation infrastructure in their communities. Within the new institutional structure of rural Kyrgyzstan, WUAs are responsible for managing the inter-farm irrigation infrastructures, while the government generally retains responsibility for the operation and maintenance of main canals. At present, there are more than 400 WUAs in Kyrgyzstan. The WUAs are expected to collect water fees, share water equitably among their members and other water users within their service area, and maintain on-farm irrigation infrastructures using fees collected from water users. However, evidence suggests that these associations are still weak and face huge problems in collecting water fees from farmers (Akramov *et al.* 2009).

3. AGRICULTURAL PERFORMANCE[52]

3.1 Changes in Land Use and Livestock

According to official classification, farms in Kyrgyzstan are currently classified into three major organizational categories: household plots, peasant farms, and corporate farms (NSC 2008). The latter are also called 'agricultural enterprises,' and include both state and cooperative (collective) farms. Household plots and peasant farms are two different types of individual (family) farm; they are differentiated largely based on their commercial orientation, size and legal status. Household plots are generally smaller and more subsistence-oriented than peasant farms, although there is some overlap between the two groups. In legal terms, household plots[53] are treated as physical entities, whereas peasant farms must be registered as legal entities. In household plots, the main farmland is a small plot of land attached to rural residence. Peasant farms operate mainly on family-owned land obtained through farm reorganization and land reform, although growth can be achieved by leasing additional land from other owners. Collective corporate farms have two main sources of land: land shares invested by individuals in the equity capital of enterprise, and the leasing of additional land from other owners.

The net effect of land reform was a dramatic shift in the distribution of agricultural land use among different farm types. The total amount of agricultural land allocated to corporate (state and collective) farms started to decline almost immediately upon the inception of transition. This trend accelerated after 1995, when the government started the widespread land reform. As shown in Figure 1, the land controlled by corporate farms dropped dramatically from about 95% of the total arable land in 1991 to 6% in 2010. Most of this land was shifted to peasant farms in the process of land reform. Currently, more than 300,000 peasant farms, with an average farmland size of 2.9 hectares, control about 84% of the total (sown) arable land in the country. The remaining 10% of total arable land is controlled by more than 900,000 traditional household plots having an average size of 0.11 hectares per holding. Another important feature of the changes in arable land use during 1990–2010 is the overall decrease in total sown area. As evident in Figure 1, between 1990

---

52 The analyses provided in this section are mainly based on official data obtained from official statistical publications of the National Statistical Committee of the Kyrgyz Republic (NSC 2008, 2012).

53 Household plots (or subsidiary household plots) are common in most transition countries of Eastern Europe and the former Soviet Union. For more information on their origins and development, see Lerman *et al.* (2004).

and 2010, the total sown arable land declined by about 150,000 hectares, which accounts for more than 11.5% of the total area sown in 1990.

Alongside the increase in land use by individual farms, the reform led to a substantial increase in livestock grazing on peasant farms and in the household sector. However, the specific pattern of the changes in livestock distribution across the different farm types differed from that seen for land use. Prior to the transition, about two-fifths of the country's cattle and about one-fifth of the sheep and goat stocks were in the care of rural households, with the remainder

Figure 1: Distribution of (Arable) Sown Area across Farm Types, 1990–2007

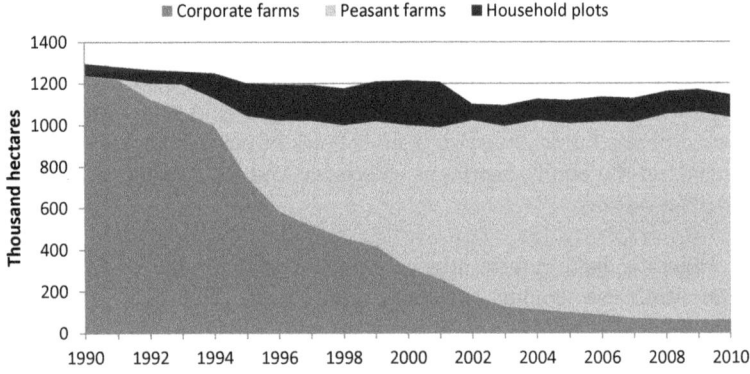

Source: NSC (2012).

held by state and collective farms. After 1991, as the large corporate (state and collective) livestock farms were privatized, the number of cattle held by corporate farms decreased radically. However, the number maintained on peasant farms and the household sector increased only moderately. As a result, by the mid-1990s, the total number of cattle in Kyrgyzstan had decreased markedly (by 30% relative to the 1990 level). Since then, the number of cattle held by households and peasant farms has increased substantially, totally offsetting the decline in the corporate farm sector (Figure 2). However, a different pattern is observed among the sheep and goat herds. The stocks held by the corporate sector declined dramatically from more than 8 million animals at the beginning of transition to less than 1 million in the mid-1990s, and eventually to about 39,000 in 2010. The number of sheep and goats held by households and peasant farms significantly increased (by more than 2.6 times) during this period. However, this increase was not enough to compensate for the decline in the corporate sector (Figure 3), yielding a dramatic decrease in the total number of sheep and goats in Kyrgyzstan (from nearly 10 million in 1990 to about 5 million in 2010).

Figure 2: Cattle Herds across Farms Types, 1990–2010

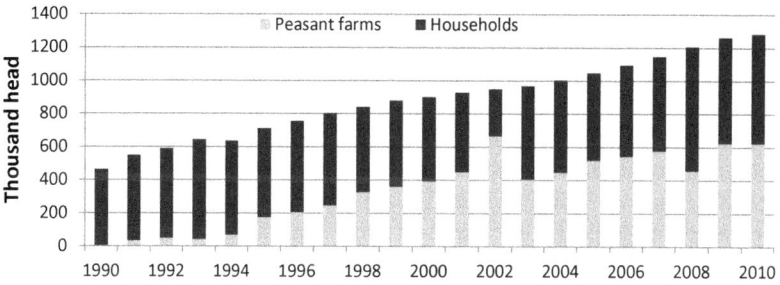

Source: NSC (2012).

Figure 3: Sheep and Goat Herds across Farms Types, 1990–2010

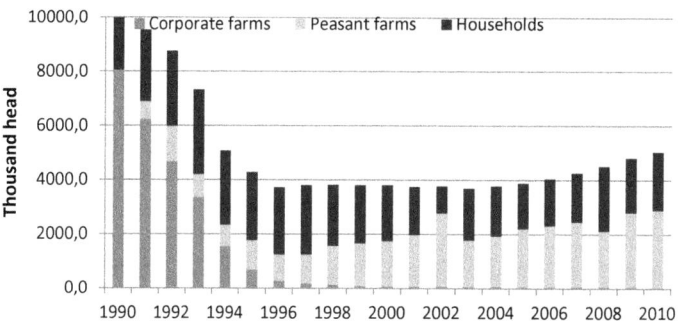

Source: NSC (2012).

The dramatic declines observed in the sheep and goat herds were largely due to shrinkage of the feed base for livestock, which forced farmers to slaughter livestock as they were unable to feed them through the winter. There are two possible explanations for the shrinkage of the feed base. First, a significant portion of the pre-transition sheep- and goat-farming sector comprised centrally funded support of fine wool sheep breeds on state and collective farms. This support helped carry the sheep herds through the winter by facilitating the importation of manufactured feed, the construction and maintenance of shelters, and the transportation of sheep (via publicly owned trucks) to and from summer pastures. After independence, the government was obviously no longer able to provide this support to sheep and goat breeders (van Veen 1995).

Second, during the transition period, the crop patterns changed to favor the production of food and high-value crops. In the Soviet economy, Kyrgyzstan's agriculture had mainly specialized in the production of livestock products, while wheat and other important food products were mainly received from outside. However, the collapse of economic relations in the former Soviet Union encouraged countries to focus on food security, leading to the increased cultivation of wheat and other food crops. As a result, the area sown to wheat, potatoes, and vegetables increased at the expense of the areas sown to barley and feed crops. Specifically, the area sown to wheat increased from 15% of the total sown area in 1991 to 32% in 2001. The land devoted to vegetables, potatoes and cotton increased from a combined 8% of the total sown land area in 1991 to 22% in 2001. In contrast, the area sown to feed crops and barley declined from about 71% of the total sown area in 1991 to 28% in 2001.

3.2 Agricultural Growth and Productivity

The changes in the distribution of arable land and livestock by farm type led to remarkable modifications in the structure of agricultural production in Kyrgyzstan. Between 1991 and 2010, the share of the individual sector (household plots plus small peasant farms) in agricultural production increased dramatically, to the point that this sector presently produces about 97% of the aggregate agricultural output in Kyrgyzstan (NSC 2008), including almost 95% of the crop production and nearly all of the livestock production. This change largely reflects the massive expansion of the peasant farms that replaced the former state and collective farms. In the crop sector, the decrease in corporate farm production was mainly offset by a corresponding rise in production by peasant farms. The phenomenon of peasant farms taking over from state and collective farms is shown with respect to grain production in Figure 4. As The decrease in grain production by the corporate sector was fully compensated by the corresponding rise in peasant farm-based production.

However, the trend is quite different for livestock production, where the individual sector was not able to compensate for the decrease in livestock production by state and collective farms. Figure 5 illustrates this phenomenon with respect to meat production, which declined dramatically (by almost 30%) during the transition period. This trend was even more dramatic in the poultry

Figure 4: Dynamic Structure of Grain Production across Farm Types, 1990–2010

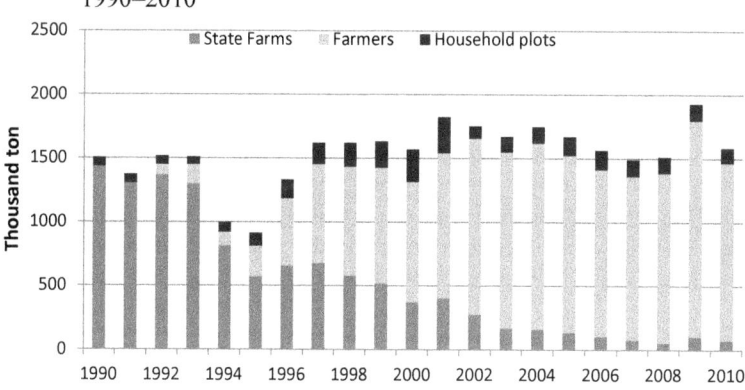

Source: NSC (2012).

and wool production, which dropped by about 50% and 70%, respectively, compared to their pre-transition levels. In contrast, the individual sector successfully compensated for the decrease in milk production by the corporate sector (Figure 6). Currently, the corporate sector plays a strictly marginal role in livestock, while the peasant farms and the household sector play equally significant roles in livestock production.

Figure 5: Dynamic Structure of Meat Production across Farm Types, 1990–2010

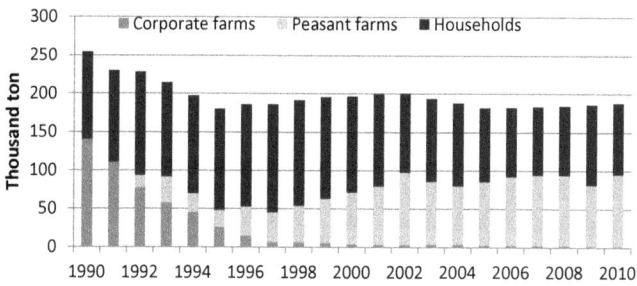

Source: NSC (2012).

The impressive shift in the composition of agricultural production does not simply reflect the dramatic changes in the composition of land holdings and livestock across different types of farms. The evidence suggests that small peasant farms are more productive[54] than collective farms, with the former producing almost two-fold more output per hectare of arable land (Table 1) and experiencing more than five-fold higher labor productivity (World Bank 2004). This could be because small farms have lower transaction costs[55] than larger farms, and/or because large farms are less efficient due to governance and agency problems. Indeed, the literature suggests that as farms become larger, costs of monitoring the production operations and enforcing labor discipline increase, eventually offsetting the gains from economies of scale (Lerman et al. 2004). Thus, the individualization of agriculture might have effectively solved the agency problems of collective agriculture and freed previously depressed private incentives, which, in turn, may have stimulated agricultural production.

Figure 6: Dynamic Structure of Milk Production across Farm Types, 1990–2010

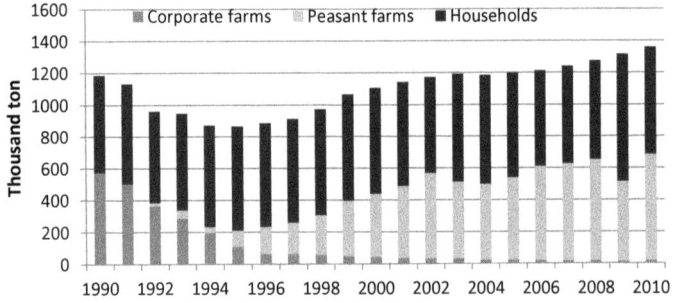

Source: NSC (2012).

Agricultural GDP (aggregate value added in the agricultural sector) and productivity growth are the ultimate measures that are often used to evaluate the overall performance this sector. Over the entire study period (1990–2008), agricultural GDP in Kyrgyzstan grew by an average of 1.4% per year. However, as shown in Table 2 and Figure 7, there were three distinct sub-periods within this agricultural growth. First, from 1991 to 1995, when the government was

---

54 However, according to official data, peasant farms appear to be significantly less (about 6-fold) productive than household plots in terms of land productivity. This is consistent with the findings of the World Bank (2004) study. At the same time, peasant farms are gradually closing the gaps with household plots, especially in terms of labor productivity. The gap in labor productivity between household plots and peasant farms is about 40% (World Bank 2004).

55 Transaction costs are the costs of running the economic system (in this case agricultural farms). They are distinguished from the production costs. Transaction costs include information, bargaining, policing and enforcement costs.

inconsistent with its agricultural policies, agricultural GDP collapsed by about 27% and showed an average negative growth rate of 6% per year. As can be seen in Figure 7, production in both the crop and livestock sectors plummeted during this sub-period. The evidence suggests that most transition countries see such production collapses, due to disproportionate increases in the prices of agricultural inputs and outputs, declining terms of trade for the agricultural sector, and dramatic declines in budgetary support for state and collective farms (Roselle and Swinnen 2004).

Table 1: Main characteristics of the farm types in Kyrgyzstan, 2007

	Household plots	Peasant farms	State and collective farms
Number, thousand	924.1	323.6	1.3
Average size of arable land holdings, hectares	0.11	2.9	58.9
Total sown area, thousand hectares	101.2	951.5	76.1
Share in total sown area, %	9	84.3	6.7
Agricultural output, million som	32931.8	53275.6	2430.8
Share in agricultural output	37	60	3
Agricultural output per hectare, thousand som	325.4	56.0	31.9

Source: NSC of Kyrgyzstan (2008; 2012) and authors' estimates.

During the second period, which coincided with serious systematic changes in land use (i.e., the liquidation of state and collective farms and the distribution of agricultural land to the members of these farms), agricultural growth rebounded after 1995 and showed a steep upward trend between 1996 and 2001. By 1999, the agricultural value added exceeded the pre-transition (1990) level, and Kyrgyzstan shifted in status from a net importer of primary agricultural products to a net exporter. Overall, agricultural GDP over this period rose by almost 60% (8% per year). This recovery of agricultural production was mainly due to shifts in the sector's institutional structure; these changes solved the incentive problems of the former collective farming system and increased the returns to labor. This augmented the labor efforts, both in terms of quantity and quality, yielding increases in labor supply and agricultural productivity.[56] Furthermore,

---

56 A similar phenomenon was observed in Chinese agriculture in the 1980s when the household responsibility system was introduced Lin (1992).

Figure 7: Gross Agricultural Output, Crops and Livestock Output, 1991–2005

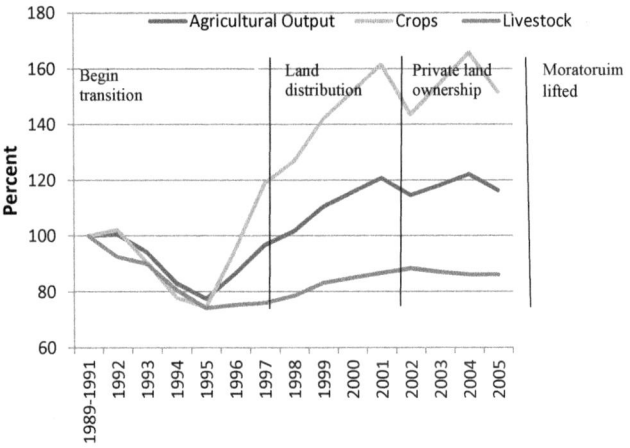

Source: FAOSTAT, NSC (2012) and authors' depiction.

by transferring decision-making power from the state to farmers, land reform allowed farmers to improve their resource allocation based on market conditions and the profitability of particular crops. This led to substantial changes in cropping patterns. The land devoted to high-value crops, such as vegetables, potatoes and cotton, increased from a combined 8% of the total sown land area in 1991 to 22% in 2001, almost a 2.5-fold increase. The change in crop patterns to favor high-value crops played an important role in improving agricultural growth.[57] In this regard, the recovery of agricultural production was mainly due

---

[57] One cannot simply claim that the entire increase in agricultural output and productivity in Kyrgyzstan between 1995 and 2001 was due to changes in farmers' incentives and the decision-making structure in the agricultural sector. The available evidence from other countries suggests that other factors, such as the liberalization of prices and input use, might also have impacted agricultural growth. China, e.g., saw a rapid average annual agricultural growth of 7.7% from 1978 to 1984 (compared with an average of 2.9% between 1952 and 1978) due to a package of market and institutional reforms that included price reform and a transition from the collective agricultural production system to an individual household- based farming (household-responsibility) system. The literature suggests that individualization of agriculture was the dominant source of the high agricultural growth rates seen in China from 1978 to1984 (Lin 1992). It was also found that the changes in agricultural prices, improvements in availability of inputs, and changes in crop patterns also had impacts on output growth. In Kyrgyzstan, however, the prices of agricultural inputs and outputs were liberalized 3-4 years before land reform, although this did not produce any positive growth under the collective agricultural

to the remarkable growth in crop production. The crop output in this period rose by 116% (13.8% per year), while livestock production rose by only 17% (2.6% per year).

Figure 8: Decollectivization (individualization) Index (DI), 1990–2007

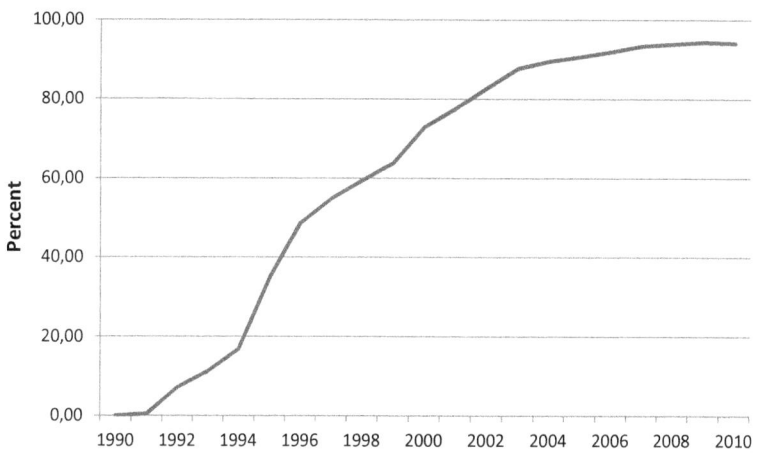

Notes: DI is calculated by dividing the difference between the share of land in individual use in total agricultural land for a given year and in 1990 by 100, minus the share of individual farms in total agricultural land in 1990: $DI=(IND_t - IND_{1989})/(100 - IND_{1989}) * 100$; see Mathijs and Swinnen (1998).
Source: Authors' calculations using data from NSC (2012).

In the third period, agricultural growth in Kyrgyzstan slowed significantly after 2002, with an average growth rate of 1.5% per year. Indeed, the growth rates of both crop and livestock production became very small. Several explanations may account for this slowdown in agricultural growth. First, the initial distribution of land from the large-scale state and collective farms to the peasant farms was completed in the early 2000s (Bloch 2002). This is shown in Figure 8, which illustrates the dynamics of the individualization of land use in Kyrgyzstan. By the early 2000s, almost 90% of the agricultural land was being used by individual farmers. Thus, it is plausible that the one-time positive

---

production system. Indeed, the evidence suggests that high input prices, limited inputs, and deteriorating governance in former state and collective farms prior to land reform led to drastic cuts in livestock numbers and falling outputs, especially in poultry products, sheep and goats (Christensen and Pomfret 2008).

discrete effect[58] of individualization of land use on agricultural growth in Kyrgyzstan ended in the early 2000s. Second, during this period, rural households often allocated labor to various non-agricultural activities, such as wage labor in urban centers within the country or seasonal labor migration abroad. Overall, from 2001 to 2007, approximately 200,000 workers exited the agricultural sector. Finally, during the same period, the arable land area sown declined by about 6%. The combination of these factors probably contributed to the observed slowdown of agricultural growth in Kyrgyzstan.[59]

Moreover, the remaining institutional and structural constraints barred peasant farmers from exploiting the full potential of private land ownership. These constraints are discussed in Section 4.

Table 2: Agricultural GDP and productivity growth in Kyrgyzstan*

	Agricultural GDP	Land productivity**	Labor productivity**
1990-2007	1.4	2.3	0.9
Sub-periods:			
1990-1995	-6.0	-4.5	-8.5
1996-2001	8.0	7.9	4.5
2002-2007	1.5	2.6	5.7

Notes: * The growth rates are annual averages for a given period.
** Land and labor productivity are defined as the average annual agricultural GDP per ha of land and per worker, respectively.
Sources: World Bank (2004), NSC (2012), and authors' estimates.

An examination of growth performance in land productivity provides a slightly different picture (Table 2 and Figure 9). Between 1990 and 2007, land productivity (measured as agricultural GDP per hectare of land) grew at 2.3% per year. Similar to agricultural GDP, land productivity declined significantly before the mid-1990s, at an average rate of -4.5% per year. While agricultural GDP increased dramatically between 1996 and 2001, the agricultural sown area did not change much. As a result, during this period, the average growth rate of the partial productivity of land was practically equal to the agricultural GDP

---

58 By 'one-time discrete impact of the individualization of land use,' we mean the behavioral and decision-making responses of peasant farmers to improved incentives resulting from the changes in the institutional setting of the agricultural sector.
59 Similar phenomena have been observed in other transition countries, such as China and Albania. Lin (1992) argued that the one-time discrete effect of the introduction of household-responsibility system ended in 1984, and this change was largely responsible for the slowing of agricultural growth in China.

growth rate. However, between 2002 and 2007, agricultural land declined while agricultural GDP continued to grow, resulting in relatively higher growth rates (2.6% per year) for land productivity. Overall, land productivity in Kyrgyzstan is relatively low at 35.9 thousand Kyrgyz soms (in 2007) per hectare of sown land area (Figure 9), which is equal to about 1000 US dollars.[60]

Figure 9: Productivity of Agricultural Labor and Land, 1991–2007 (in constant 2007 prices)

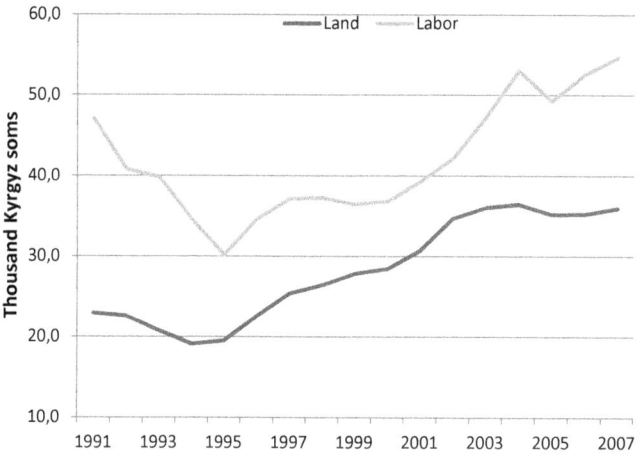

Source: NSC (2012) and author's estimates.

In terms of labor productivity, the average worker employed in the agricultural sector in 2007 produced an agricultural value added of about 54.6 thousand Kyrgyz soms (measured in constant 2007 prices); this was approximately 16%, 81%, and 39% higher than the levels seen in 1991, 1995, and 2001, respectively (Figure 9). The pattern of labor productivity[61] growth was considerably different from the growth patterns seen for agricultural GDP and land productivity. In the early 1990s, the agricultural sector absorbed the unemployed population, which had increased significantly due to the collapse of the industrial sector in the beginning of transition. This coincided with substantial declines in sown area and agricultural output (between 1991 and 1995). The combined result of these trends was a large decline (on average, -8.5% per year) in labor productivity. As

---

60 Kyrgyzstan's average annual exchange rate in 2007 was equal to 37.3085 Kyrgyz soms per US dollar.
61 Labor productivity is measured as the value of agricultural GDP per worker employed in the agricultural sector.

a result, in 1995, the level of labor productivity in the agricultural sector was about 60% of the pre-transition (1990) level.

Decollectivization and the distribution of land to members of former state and collective farms attracted even more labor to the agricultural sector from other sectors of the economy; total employment in the sector reached almost 950,000 persons in 2001, which was about 50% higher than the pre-transition level (Figure 10). Despite the massive increase in the agricultural workforce and the

Figure 10: Agricultural GDP, Land, and Labor, 1990–2007 (1990=100)

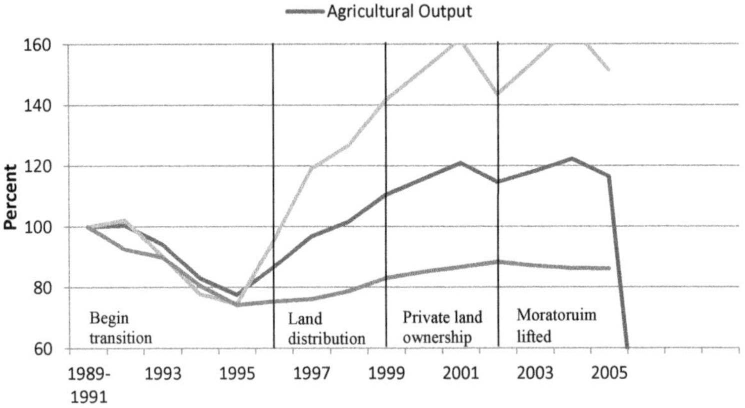

Source: NSC (2012) and authors' estimates.

relatively stable sown area, labor productivity rebounded due to the higher growth rates in agricultural GDP. The average growth rate of labor productivity between 1996 and 2001 was 4.5% per year. After 2002, sown arable land area declined, the trends in agricultural employment reversed, and labor started to move out of agriculture to other sectors of the economy. Although agriculture still remains the main productive activity undertaken by most households in rural Kyrgyzstan, many households augment their agricultural incomes with a wide array of other productive activities. At present, rural households frequently allocate labor to various self-employment activities, to wage labor in urban centers within the country, or to migration. With the recovery of the industrial sector and construction in Russia and Kazakhstan, demand for seasonal labor has increased. Many rural households in Kyrgyzstan allocate labor to seasonal migration, because the economic returns from seasonal migration are often significantly higher than those from agriculture[62] (ADB 2008). During this period, agricultural employment and sown area declined by about 22% and 6%, respectively (Figure 10). The combined effect of these trends was a moderate increase in the land/labor ratio from 1.21 hectares per worker in 2002 to 1.52 hectares in 2007 (the latter is still 26% below the 1991 level). As a result, despite a relatively small increase in agricultural GDP, the partial productivity of labor over the study period increased by about 5.7% per year.

4. CONSTRAINTS ON AGRICULTURAL GROWTH

There is a certain consensus among economists that better property rights institutions lead to better economic outcomes. The literature considers three channels for this positive link (Deininger and Feder 2001; Do and Iyer 2008). First, stable property rights provide greater incentives for labor efforts and long-term investments in land and new agricultural technology. A second channel is through the enhanced possibilities for transfer of land to more efficient users. Third, if better property rights make it easier to use land as collateral, then agricultural growth might be improved through better access to credit and reduced constraints on investment. Our examination of agricultural performance in the previous section showed that land reform in Kyrgyzstan played an important role in the recovery of agricultural output, resulting in remarkable growth between 1995 and 2001. The shift in the sector's institutions created greater incentives for additional labor efforts, leading to the recovery of agricultural production and growth. However, remaining problems with inadequate institutional environment have not allowed private farmers to fully realize the benefits of private land ownership to date. In the following, we discuss the remaining institutional constraints limiting agricultural growth in rural Kyrgyzstan. This analysis mainly uses secondary data from a recent farm survey conducted by the M` Vector Consulting Company based in Bishkek for German Society for Technical Cooperation (GTZ).

4.1 Fragmentation of Farmland

One negative impact of land reform in Kyrgyzstan was its contribution to the fragmentation of land use and its impact on collective action in the rural economy. The distribution of agricultural land based on universal and equitable principles inevitably fragmented land holdings in the country, creating hundreds of thousands of small peasant farms. The average size of the land holdings for

---

62 Currently, Kyrgyzstan is one of the world's largest recipients of international remittances. According to a recent World Bank-commissioned study, remittances measured in the balance of payments constitute about 10% of Kyrgyzstan's GDP (Quillin et al. 2007).

peasant farms is about 2.9 hectares, with about 80% of Kyrgyzstan's individual farmers' having farmland area less than 2 hectares (World Bank 2007). A recent survey conducted by a local consulting company in Kyrgyzstan showed that almost half of the individual farmers use less than 1 hectare of arable land, while another 40% use between 1 and 3 hectares (M`Vector Consulting 2007). Land fragmentation is an even more serious issue in the southern provinces of the country, which have higher population densities. For example, in Osh province, about 70% of farmers' land holdings are less than 1 hectare (M`Vector Consulting 2007).

The fragmentation of agricultural land has important implications for agricultural productivity and growth. The relationship between farm size and labor productivity has not been clearly established in the literature. One popular stylized fact suggests that small farms are more productive per unit of land than large farms. This inverse relationship between farm size and land productivity is usually explained by differences in factor endowments between small and large farms (Ellis 1993; Fan and Chan-Kang 2005). However, the literature also suggests that the relationship between farm size and productivity depends on opposing cost economies and diseconomies for a given technology (Ellis 1993). Cost economies might result from the indivisibility of a fixed capital, labor divisions, and marketing economies in the bulk of input purchases or output sales. Cost diseconomies relate to the limits of effective management and labor supervision, and the changing nature of risks as the farm scale increases. With the transformation of agriculture toward a science-based approach, labor becomes less important in shaping land productivity, while other inputs such as machines, fertilizer, and irrigation play greater roles (Ellis 1993).

The potential impact of the decollectivization of agriculture and the fragmentation of land use on productivity depends on which prevail: cost economies or cost diseconomies. The strong growth in Kyrgyzstan's agricultural sector since the initiation of land reform suggests that any negative impact has been offset by improvements in labor efforts and resource allocation. However, in the presence of inadequate institutions, imperfect markets for land, credit, input and output, and deficiencies in property rights, fragmentation may hamper the expected long-term benefits of land reform and limit future growth in the sector.

The evidence from other transition economies suggests that, in the individual sector, peasant farms cultivating larger holdings are more efficient and produce relatively higher family incomes compared to farms with relatively smaller allotments (Lerman 2002). Since the distribution of land in the 1990s, Kyrgyzstan has developed appropriate institutional and legal arrangements for the development of a land market, including the establishment of a system for registering property rights. In 1999, the State Agency for Registration of Rights (Gosregister) was established, along with a nationwide network of local registration offices. In 2001, the moratorium on agricultural land sales was lifted. Nevertheless, the evidence suggests that legal and policy impediments are still restricting the development of an efficient agricultural land market (USAID 2008). As mentioned in Section 1, the law states that only citizens of Kyrgyzstan, who are residents of rural areas and at least 18 years old, residing in a given rural community for no less than two years, can own agricultural land. This requirement clearly restricts activity in the market by limiting the pool of potential buyers of agricultural land. However, land consolidation cannot be limited to the buying and selling of agricultural land. The experience of market economies shows that leasing can play an important role in enlarging the size of landholdings. As Lerman (2004) noted, leasing has emerged as an important mechanism for the enlargement of individual farms' landholdings in transition economies, such as those in Moldova, Hungary and Poland.

In Kyrgyzstan, the law permits the lease of agricultural land, and market-driven land consolidation through leasing has already begun. According to USAID (2008), approximately 50% of corporate and peasant farmers lease land from others, including the lease of LRF land from local governments (indeed, most of these are probably leases of LRF land). The findings from a recent farm-level survey suggested that about 10% of farmers lease additional agricultural land from other private land owners (M'Vector Consulting 2007). The evidence also suggests that short-term term land lease agreements, especially between private land owners and individual farmers, prevail in Kyrgyzstan (USAID 2008). This could be a serious obstacle to land consolidation and agricultural growth, as it discourages long-term investment by lessees. Government policy, therefore, should encourage longer-term leasing. However, formal government-sponsored land consolidation programs should not be based on coerced cooperation; rather they should augment the market-driven process that has already begun.

4.2 Access to Agricultural Services

The lack of adequate institutions and markets can create significant problems for small peasant farmers without access to machinery and traditional agricultural services. During pre-transition times, all state and collective farms had their own units and specialists responsible for the delivery of such services. Due to the limited size of small peasant farms, however, it is difficult (if not impossible) for them to support such units or purchase machinery. In addition, individual farmers may need other services, such as marketing, legal, and extension services, to fully exploit the new economic paradigm of the market economy.

Unfortunately, the agrarian reforms in Kyrgyzstan were inconsistent, and this dimension of transformation to private farming was mostly neglected during the earlier phases of transition. For example, with the reorganization of large farms into small individual farms, there was little or no linkage between the new farms and agricultural research due to the lack of extension services (Schmidt 2001).

Only in the late 1990s, with support from external donors, the government started to emphasize the construction of an institutional and market environment for a market-based agricultural sector. This included the initiation of several programs aimed at creating institutions to provide agricultural support services for farmers. Such institutions included: the Rural Advisory Service (RADS), which provides advisory (extension) services to peasant farmers; the Kyrgyzstan Agricultural Market Information Services (KAMIS), which supports farmers by providing market and price information; and Legal Assistance to Rural Citizens (LARC), a donor-supported institution that provides legal services to rural citizens. Currently, these institutions have offices in most rural districts. However, only few farmers appear to be aware of these institutions, and only a fraction of them actually use their services (M`Vector Consulting 2007):

(i) About 20% of farmers are aware of the RADS and only 16% use its advisory (extension) services.
(ii) Less than 10% of farmers are aware of KAMIS and less than 2% actually use its marketing and information services.
(iii) Only 7% of farmers know about LARC and less than 3% use its legal services.

Furthermore, both economic theory and the experiences of market economies suggest that small farmers can make use of scale economies by establishing service cooperatives (Lerman *et al.* 2004). This strategy creates a cooperative machinery pool and allows small farmers to share the burden of capital expenditures. The literature suggests that there is strong psychological resistance to cooperation in transition economies, due to years of abuse of the concept of 'cooperation' prior to transition (Lerman 2004). Nevertheless, the experience of some transition countries suggests that, given appropriate legal and institutional arrangements, cooperation might significantly improve the access to machinery and agricultural services (Lerman *et al.* 2004). This is especially important for Kyrgyzstan, where only about 50% of peasant farmers use agricultural machinery services. The appropriate legal basis for the development of such cooperatives was finally created in Kyrgyzstan by the adoption of the 'Law on

Trade and Service Cooperatives.'[63] The above mentioned farm survey (Survey 2007) found that many farmers appear to be keen to join service cooperatives. While only 1% of farmers reported that they were already members of a service cooperative, an additional 44% indicated that they would be willing to join a service cooperative if such an institution were established in their community. Thus, promoting the establishment of trade and service cooperatives may be a promising direction for fostering improvements in farmers' access to agricultural machinery and services.

Another important problem is related to collective action in the delivery of and access to irrigation water, which has been drastically affected by the fragmentation of land use. As mentioned in Section 2.3, following the land reform, the government started to reform the irrigation sector through establishment of WUAs. However, institutional reform in the irrigation sector was significantly slower than land reform, and the facilities were already largely deteriorated by the time reforms were initiated in the irrigation sector. There are currently about 400 WUAs in Kyrgyzstan; they are responsible for the operation and maintenance of on-farm irrigation facilities and the allocation of irrigation water to peasant farmers in most parts of the country. However, empirical evidence suggests that the WUAs are still weak and financially unsustainable, due to problems with the collection of irrigation service fees (Akramov *et al.* 2009, Johnson and Stoutjesdijk 2008). Moreover, merely changing a regime and creating laws seem insufficient for successful institutional change. Theesfeld (2004) argued that important societal features inherited from socialism and amplified during the transition process, such as incongruities between formal and informal rules, power abuse, rent seeking, information asymmetry, and distrust between community members, played a crucial role in evolution of social capital and collective action in Bulgaria's irrigation sector. The volume of anecdotal evidences encountered by the authors of the present paper suggests that similar problems related to collective action are applicable to Kyrgyzstan's irrigation sector. Problems with information asymmetry and inadequate access to information seem to lead to insufficient or delayed water access for peasant farmers in Kyrgyzstan. Thus, it is not surprising that about 36% of peasant farmers reported that access to irrigation water is the most important problem in the agricultural sector (M'Vector Consulting 2007).

---

63 The draft version of this law was developed with technical assistance from the German Technical Cooperation Agency (GTZ).

4.3 Access to Credit

The impact of land privatization on the access to credit for Kyrgyzstan's small farmers, as elsewhere, is subject to two important constraints (at least in the short and medium terms). The first constraint is high transaction costs to financial institutions when dealing with small-scale borrowers. Obviously, loan processing usually involves a strong scale of economy, making it costly to collect information on the past behavior of small farmers and to assess the potential profitability of their small projects. This is perhaps one of the reasons why commercial banks in Kyrgyzstan often refuse to take small land plots as collateral. In fact, the evidence suggests that the size of farmland has a significant impact on the likelihood of farmers' credit access in Kyrgyzstan. Farmers with landholdings of less than 1 hectare and 1-3 hectare are about 3-fold and 1.6-fold less likely to use credit compared to farmers with landholdings of more than 3 hectare (Akramov and Omuraliev 2009). Furthermore, under existing law, commercial banks cannot own agricultural land, and they have only a limited right to take a possession of pledged land in foreclosure (USAID 2008). By limiting agricultural land ownership only to residents of rural areas residing in a given rural community for no less than two years, the law restricts the market value of agricultural land. Property rights to land can enhance the collateral value of land only if the lender is able to gain full possession of mortgaged land in case of default.

The second constraint is the overall shortage of credit supply in Kyrgyzstan.[64] In general, the country's formal financial sector is very small, especially in rural areas. The total credit to the economy from banks and non-bank financial institutions comprises about 20% of GDP. Moreover, the broad money supply is comprised mostly of currency outside banks (about 70% of the money supply), with deposits in the financial sector comprising only 30% of the money supply. This low level of financial intermediation is usually attributed to a general lack of confidence in the banking system, stemming from various banking crises that occurred during transition. The absence of banks and their branches in rural areas is also an important component of the relative lack of credit. Among the commercial banks, only Ayil (village) Bank has a considerable involvement in rural financial markets. As a result, agricultural credit consists of only 12% of total bank credits in Kyrgyzstan, while the sector employs about 40% of the country's labor force and contributes about one third of the country's GDP. The government and donors (e.g., the World Bank, ADB and USAID) are currently promoting credit unions and microfinance institutions in rural areas, to the point

---

64 The analysis in this paragraph is based on data obtained from the official website of the National Bank of the Kyrgyz Republic: www.nbkr.kg (accessed on 15 January 2010).

there are now some 270 credit unions and 230 microfinance institutions available in the country. The evidence suggests that these non-bank financial institutions are starting to play an important role in expanding access to credit in rural Kyrgyzstan, as they have done in so many other countries. Nearly 40% of farmers that use credit to finance their farming activities receive the credit from such financial institutions (Akramov and Omuraliev 2009).

5. CONCLUSION

The major agricultural reforms in Kyrgyzstan during transition consisted of the complete liquidation of former large-scale state and collective farms and the distribution of land to rural households. This established private property rights to land, with rights to transfer, exchange, sell, lease and use the land as collateral for credit. In addition, reforms in the irrigation sector established an institutional and legal environment in which collective action management solutions may be implemented based on water users associations. These key features of Kyrgyzstan's reform are in sharp contrast with those seen in other transition economies in Central Asia, where decollectivization was not complete and the reforms were implemented gradually.

The institutional modifications in Kyrgyzstan led to remarkable changes in the structure of agricultural production, and the individual sector now produces about 97% of the country's agricultural output. Coinciding with these changes, growth in agricultural output and productivity rebounded, showing steep upward trends between 1996 and 2001. This remarkable outcome was the result of the land reform, which augmented labor efforts and triggered substantial changes in cropping patterns. However, although labor productivity continued to grow by a respectable 5.5% per year after this point, the growth rates of agricultural GDP and land productivity slowed after 2002. This was probably due, at least in part, to the combination of three factors: the end of the one-time positive discrete effect of the individualization of land use; an exodus of labor from the agricultural sector; and a slight decrease in sown area.

Some argued that significant declines in agricultural growth are associated with excessive fragmentation of land use and the apparent inability of small peasant farmers to sustain growth in agricultural productivity (Government of Kyrgyzstan 2004). However, our analysis of official data suggests that this claim is inaccurate, given that the small peasant farms produce almost 2-fold more agricultural output per hectare of sown area than the larger corporate farms in Kyrgyzstan. Thus, we argue that the remaining constraints associated with inadequate institutional and market environment may have contributed to the slowing of agricultural growth. In order for farmers to fully realize the benefits of their improved property rights, policymakers should remove the remaining institutional and legal barriers for land consolidation and access to bank credit, promote collective action and the expansion of producer organizations, trade and service cooperatives, and help improve farmers' access to agricultural services.

There is very little evidence in the literature examining how institutional and structural changes in rural Kyrgyzstan have affected the country's agricultural production and productivity. In the future, research that further illuminates the role of these changes in advancing agricultural productivity could potentially help the policy debate in Kyrgyzstan. In this regard, empirical analysis based on province- and district-level panel data, seeking to identify the differences and variations in agricultural productivity across sub-national jurisdictions, will be interesting. Additionally, studies that further explore institutional reform in irrigation water management and the effects of water users associations on agricultural productivity may be fruitful areas of research.

REFERENCES

Akramov, K. and Omuraliev, N. (2009), 'Institutional change, rural services and agricultural performance in Kyrgyzstan', *International Food Policy Research Institute Discussion Paper 904.*

Akramov, K., Crewett, W. and Omuraliev, N. (2009), 'Institutional reform in irrigation water management and agricultural productivity in Kyrgyzstan', International Food Policy Research Institute.

Anderson, K.H. and Pomfret, R. (2003), *Consequences of creating a market economy: Evidence from household surveys in Central Asia*, Northampton, MA: Edward Elgar Publishing.

Asian Development Bank (ADB) (2008), 'Remittances and poverty in Central Asia and South Caucasus', *Technical Assistance Consultant's Report Number 40038.*

Bloch, P.C. (2002), 'Land reform in Kyrgyzstan: Almost done, what next?' *Problems of Post – Communism 49(1): pp. 53-62.*

Bloch, P.C., Delehanty, J.M. and Roth, M.J. (1996), 'Land and agrarian reforms in the Kyrgyz Republic', University of Wisconsin-Madison Land Tenure Centre Research Paper 128.

Christensen, G. and Pomfret, R. (2008), 'The Kyrgyz Republic', in: Anderson, K. and Swinnen, J. (eds.), *Distortions to agricultural incentives in Europe's transition economies* Washington, D.C.: World Bank, pp. 265–296.

Deininger, K. (2002), 'Agrarian reforms in Eastern European countries: Lessons from international experience', *Journal of International Development*, 14, 7, pp. 987–1003.

Deininger, K. and Feder, G. (2001), 'Land institutions and land markets', in: Gardner, B.L. and Rausser, G.C. (eds.), *Handbook of agricultural economics (Volume 1)*, Amsterdam: Elsevier, pp. 288–331. Do, Q.T. and Iyer, L. (2008), 'Land titling and rural transition in Vietnam', *Economic Development and Cultural Change,* 56, 3, pp. 531–580.

Ellis, F. (1993), *Peasant economics: Farm Households and Agrarian Development*, Cambridge, UK: Cambridge University Press.
Fan, S. and Chan-Kang, C. (2005), 'Is small beautiful? Farm size, productivity and poverty in Asian agriculture', *Agricultural Economics*, 32, 1, pp. 135–146.
Gioveralli, R. (1998), 'Land reform and farm reorganization in the Kyrgyz Republic', Rural Development Institute Report on Foreign Aid and Development No. 96.
Government of the Kyrgyz Republic (2004), 'Agrarian policy concept of the Kyrgyz Republic to 2010', *Approved by the Resolution of the Government of the Kyrgyz Republic No. 465 on 22 June 2004.*
Johnson III, S. and Stoutjesdijk, J. (2008), 'WUA training and support in the Kyrgyz Republic', *Irrigation and Drainage*, 57, 3, pp. 311–321.
Lerman, Z., Csaki, C. and Feder, G. (2004), *Agriculture in transition: Land policies and evolving farm structures in post-Soviet countries*, Oxford, U.K.: Lexington Books.
Lerman, Z. (2002), 'Productivity and efficiency of individual farms in Poland: a Case for land consolidation', *Paper presented at the Annual Meeting of the American Agricultural Economics Association*, Long Beach, CA, 28–31 July 2001.
Lerman, Z. (2004), 'Policies and institutions for commercialization of subsistence farms in transition countries', *Journal of Asian Economics*, 15, 3, pp. 461–479.
Lerman, Z. (2008), 'Agricultural development in Central Asia: A survey of Uzbekistan', *Eurasian Geography and Economics*, 49, 4, pp. 481–505.
Lin, J.Y. (1992), 'Rural reforms and agricultural growth in China', *The American Economic Review*, 82, 1, pp. 34–51.
Macours, K. and Swinnen, J. (2002), 'Patterns of agrarian transition', *Economic Development and Cultural Change*, 50, 2, pp. 365–394.
Mathijs, E. and Swinnen, J. (1998), 'The economics of agricultural decollectivization in East Central Europe and the former Soviet Union', *Economic Development and Cultural Change*, 47, 1, pp. 1–26.
Murrell, P. (2008), 'Institutions and transition', in: Durlauf, S.N. and Blume, L.E. (eds.) *The new Palgrave dictionary of economics*, New York: Palgrave Macmillan.
M' Vector Consulting (2007), *Farm survey on readiness and motives of farmers to join cooperatives*, Bishkek, Kyrgyzstan.
National Statistical Committee (NSC) of the Kyrgyz Republic (2008), 'Agricultural sector of the Kyrgyz Republic: 2003-2007', *Statistical Publication*, Bishkek, Kyrgyzstan: NSC (in Russian).
National Statistical Committee (NSC) of the Kyrgyz Republic (2012), 'Agricultural sector of the Kyrgyz Republic: 1991-2007', *Dynamic Statistical Tables*, Bishkek, Kyrgyzstan, online at http://www.stat.kg/rus/part/agri.htm.
Quillin, B., Segni, C., Sirtaine, S. and Skamnelos, I. (2007), 'Remittances in CIS countries: A study of selected corridors', *World Bank ECSPF Regional Working Paper 2*.
Roselle, S. and Swinnen, J.F.M. (2004), 'Success and failure of reform: Insights from the transition of agriculture', *Journal of Economic Literature*, 42, 2, pp. 404–456.
Sarris, A., Doucha, T. and Mathijs, E. (1999), 'Agricultural restructuring in Central and Eastern Europe: Implications for competitiveness and rural development', *European Review of Agricultural Economics* 26, 3, pp. 305–329.
Schmidt, P. (2001), 'The scientific world and the farmer's reality: Agricultural research and extension in Kyrgyzstan', *Mountain Research and Development*, 21, 2, pp. 109–112.
Spoor, M. (2003), *Transition, institutions, and the rural sector*, Oxford, U.K.: Lexington Books.

Theesfeld, I. (2004), Constraints on collective action in a transitional economy: The case of Bulgaria's irrigation sector, *World Development*, 32, 2, pp. 251–271.

USAID (2008), *Land market development in Kyrgyzstan: Analysis and recommendations.* Washington, D.C.: USAID.

Veen, T. van (1995), 'The Kyrgyz sheep herders at a crossroads', *Pastoral Development Network Series* 38.

World Bank (2004), *Sustaining pro-poor growth: Emerging challenges for government and donors - Kyrgyz Republic agricultural policy update*, Washington, D.C.: World Bank.

World Bank (2007), *Kyrgyz Republic poverty assessment: Growth, employment, and poverty.* Washington, D.C.: World Bank.

Factors Determining Crop Insurance Market Development in a Transition Economy: The case of Kazakhstan

Olaf Heidelbach and Raushan Bokusheva

1. INTRODUCTION

Kazakhstan's agricultural sector plays a vital role in the country's economy. It does not only function as an economic output producer, it also serves as a social buffer in the country's ongoing transition to a market economy. The restructuring process has had a strong impact on the economic performance of agricultural enterprises. As the state no longer functions as a back-up financier in times of economic downturn, farmers have had to find their own sustainable instruments for managing business risks, which are significant in Kazakhstan due to the acute continental climate and the resulting revenue fluctuations in crop production. Besides *on-farm strategies* that concern farm management and include production portfolio selection, holding sufficient liquidity and diversification, *risk-sharing strategies* can play an important role in agricultural income stabilization. Risk-sharing strategies include marketing contracts, production contracts, vertical integration, hedging on futures markets, participation in mutual funds and insurance.

This contribution attempts to shed light on factors influencing the development of the crop insurance market in Kazakhstan. Three objectives will be pursued: First, factors influencing the development of the crop insurance market, both from the demand- and supply-side will be identified. Second, crop insurance's capacity to efficiently stabilize income under transition circumstances will be evaluated. In the final step, recommendations for developing crop insurance as an effective income stabilization tool will be made to political decision-makers.

2. PROBLEMS OF AGRICULTURAL INCOME STABILIZATION AND INSTITUTIONAL HISTORY OF CROP INSURANCE

In the continental-climatic vegetation conditions of Kazakhstan, plant production carries a particularly high risk, which manifests itself predominantly in the considerable volatility of yields. Comparing major grain producers' coefficients of variation illustrates Kazakhstan's relative comparative disadvantage in rain-fed crop production. For wheat as well as for sunflowers, Kazakhstan ranks last in the list of most important producer countries. In both cases, Kazakhstan produces the lowest yields, with a standard deviation comparable to other transition countries (Russia, Ukraine) that produce double the mean yield. Figure 1 depicts mean values and variation coefficients for wheat yields in selected countries.

Figure 1: Mean and variation of national wheat yields (1980–2003)

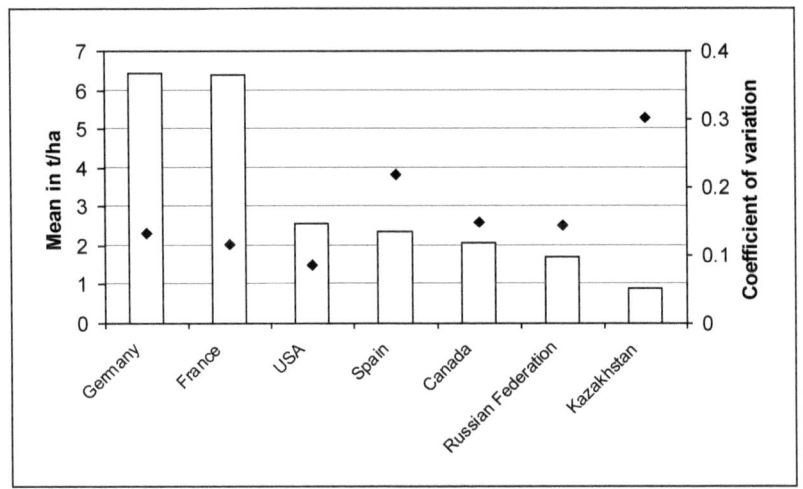

Note: Data from 1980–2003.
Sources: FAOSTAT, Statistical Yearbook USSR (1985), Statistical Handbook States of the Former USSR (1995), Statistical Yearbook CIS countries (1994), Statistical Yearbook Kazakhstan (1984, 1987, 1990, 2004).

Crop insurance has a long history in the former USSR, both in general and in Kazakhstan particularly. Indeed, agricultural insurance has existed since tsarist times, but one of the characteristics of the socialist crop insurance system was a relatively rough estimation of insurance tariffs, i.e., there was not much differentiation between regions within the Soviet republics.

After the dissolution of the Soviet Union in 1991, the state insurance system was suspended. Five years later, a new mandatory crop insurance system was introduced under the supervision of the state insurance company "Kazagropolis". However, "Kazagropolis" went bankrupt because many farmers could not pay their premiums (Uaisovich 2003). On the other hand, farmers interviewed in the framework of a farm survey stated that the state insurance company did not pay indemnities in case of a crop loss. In 2001, "Kazagropolis" was closed down, but some of the negative image of the former state insurance system remained. This bad reputation, the result of past failures, is clearly an obstacle to the new crop insurance system, which is based on private insurance industry participation.

Under the national agro-food program of the Republic of Kazakhstan, the current law "On Mandatory Crop Insurance" was adopted in 2004. However, due to the absence of insurance companies that hold a license for crop insurance activities, implementing the law in 2004 was impossible. Practical implementation started in 2005 with the beginning of the sowing campaign. The main objective of the mandatory crop insurance law is to protect, through insurance payments, crop producers' property from the consequences of adverse natural phenomena which might lead to partial or complete loss of harvest. The system's main characteristics are fixed premiums (depending on the respective crop and region) and the indemnification of production costs, subject to the technology used at a farm.[65] Since the law has been adopted, the dissemination process has provoked mostly depreciative reactions from farmers and farmers' organizations. Contrary to insurance representatives' arguments, (Tazhmakin cited in Prokhorov 2005), it is not only poor farmers that disapprove of crop insurance. According to experience gained during extensive visits to the country, there is broad distrust of the current crop insurance system. International experience shows that large farmers and the insurance industry often profit the most from crop insurance programs.

3. INSURANCE SUPPLY

In the framework of the new law on crop insurance, licenses *without time limit* for insurance activities have been issued to four insurance companies (JSC Grain Insurance Company, JSC TransOil Insurance Company, JSC Victoria Insurance Company, and JSC Eurasia Insurance Company). Initially, active work in concluding contracts was carried out by the former two companies, only. However, the activities of these two insurance companies turned out to be insufficient for providing complete mandatory crop insurance coverage. In 2004, the total insured area in Kazakhstan was 8,225,998 ha, which is 57.6% of the entire crop area subject to mandatory insurance (14,278,000 ha). There is a higher penetration of crop insurance in regions with a higher share of large-scale enterprises and better infrastructure, such as Akmola, Kostanai, East, and North Kazakhstan. Smaller private farms have only restricted access to insurance, even though they would need it more than large farms due to limited self-insuring opportunities. According to the statements of the then-chairman of the *Grain Insurance Company*, a company whose main interest lies in insuring large-scale

---

[65] There are three types of technologies ("scientifically-justified", "standard", and "extensive") that determine indemnity levels.

farms in northern Kazakhstan, this development was predictable (Tazhmakin 2003).

In 2005, during the course of practically applying the law's norms, many shortcomings were discovered in the newly-introduced crop insurance system. Thereafter, a working group within the Ministry of Agriculture identified three main problems of the current system:

- *Monitoring*: The monitoring committee's authority is questioned by farmers and insurance companies; distances between plots are large and infrastructure underdeveloped, preventing qualified monitoring from being achieved on time.
- *Communication*: Insufficient information about the insured's characteristics (production volume, applied technology) caused by missing collaboration between agricultural administration, statistical offices and insurance companies. Different, sometimes shady practices for transferring premiums and indemnities.
- *Inadequate provision of information* to farmers and regional agricultural administration regarding insurance mechanism and regulation: this leads to confusion and resentment among farmers in different regions, e.g. productive farmers in northern Kazakhstan do not want to subsidize highly risk-prone farms in western Kazakhstan.

Further weak points of the insurance program can be derived by studying the institutional framework and natural conditions in Kazakhstan:

- *Moral hazard* may arise, as farmers could select the highest coverage (corresponding with scientifically-justified technology), although they have used simplified technologies in the hope of increasing the probability of insurance payout.
- The law does not make a detailed distinction between low-risk areas and high-risk areas. This might expose the whole system to *adverse selection*, particularly for crops other than cereals that have a single country-wide premium rate.
- *A slow processing rate and a high claim rejection rate*. The loss adjustment is performed by a commission that determines yield loss. The first year processing rate was very low due to implementation difficulties.

In spite of many implementation difficulties, in 2006 the number of insurance contracts and the total insured area increased from the previous year in nearly all oblasts (Appendix A). Most farms buy insurance contracts only to avoid fines, which can be much higher than insurance premiums. Additionally, insurance penetration continues to be very low in southern Kazakhstan oblasts (South Kazakhstan, Almatinskaya, Djambulskaya, Kysylordinskaya), where small farms play a predominant role.

Another evident problem on the way to a functioning commercial crop insurance system is the *lack of interest of well-established insurance companies*. Insurer liquidity problems can be caused by systemic risk and credit-rationing. Credit-rationing could be overcome by diversifying the insurer's portfolio with insurance types where indemnity payments are not correlated with those of crop insurance. In Kazakhstan, the insurance market is not yet fully developed, and one might question the adequacy of relatively small companies acting as a national crop insurer.[66]

Private insurance companies clearly act as profit-maximizers by keeping transaction costs low. Interviews with selected representatives of insurance companies shed light on their attitude towards crop insurance. The interviewees mentioned several reasons for private insurers' low interest in crop insurance:

- functionality of the scheme is highly dependent on state budget;
- non-liquidity of farms (and impossibility of choosing clients);
- lacking insurance methodology (actuarial soundness);
- superficial calculation of premiums (how to monitor production costs?);
- reinsurance difficult to attain with presence of systemic risk;
- interests of the insurance industry were not considered when formulating the law;
- insufficient and expensive access to information (e.g. weather and yield data);
- monitoring (asymmetric information problem) is not sufficiently solved by the law (no trust in the monitoring commissions).

Some of these problems could be solved by introducing index-based insurance. These products, however, were not widely tested in practice, and an introduction on a pilot-basis might be prohibitively expensive and thus not feasible for private insurance companies. Therefore, government or international donors should provide necessary expertise and financial means.

---

[66] The three actively engaged insurance companies own considerably less than average capital compared to the entire sample of 37 insurance companies in Kazakhstan (National Bank 2006). Own capital equipment is particularly important when systemic risk prevails.

Keeping in mind the described implementation problems on the supply-side, the next chapter presents binary choice model results on factors influencing insurance demand, and also tests the feasibility of index-based insurance for Kazakhstan's agriculture.

4. DEMAND FOR CROP INSURANCE IN KAZAKHSTAN

In this section we evaluate demand for crop insurance from two perspectives – descriptive and normative. In the first part, by applying a logit-model we determine factors that influence farmers' decisions to purchase crop insurance; subsequently, by means of a normative programming model, we analyze whether and under which conditions crop insurance can become an effective risk-reducing instrument for farmers in Kazakhstan. The descriptive analysis is based on survey data from 73 farms from 6 regions in Kazakhstan. The applied utility-efficient programming model is specified for two typical farms in northern and eastern Kazakhstan.

Factors influencing the decision to purchase crop insurance

Three studies have estimated models of crop insurance purchase decisions using county level data (Smith and Baquet 1996; Smith and Goodwin 1996; Coble et al. 1996). Van Asseldonk et al. (2002) as well as Mishra and Goodwin (2006) investigate factors influencing the demand for crop insurance products using farm-level and household data. This approach seems promising from a methodological viewpoint. Thus, the present study uses farm-level data collected in the framework of a farm-survey among 73 farms from six regions in Kazakhstan and employs a binominal logit model to identify the main factors determining crop insurance demand.[67]

The results in Table 1 include the description of all variables, the values of the coefficient estimates, the t-statistics and general statistics to verify the overall model. The findings are largely consistent with anticipated relationships and indicate that potential crop insurance buyers are characterized by a specialization in grain production and a high degree of concern about income losses as a consequence of risk. Specialization on grain production increases income variability, and thereby the need for income-stabilizing instruments. Diversified farms, e.g. in southern Kazakhstan, have a lower crop insurance demand. Potential crop insurance service buyers can be found in the group of respondents concerned about income loss as a consequence of risk. Neither

---

67 The model was tested for multicollinearity. All variables that showed multicollinearity were removed from the model.

concerns about insolvency as an outcome of risky business, nor farm size and past experiences with crop insurance are of any significance for the decision to participate.

The adoption of insurance (and technology) is grounded to some extent in human capital theory. Welch (1970), Khaldi (1975), Nelson (1985) and Wozniak (1989) use education as a measure of human capital to reflect the ability to adopt innovation (either technology or insurance). The education level is proven to be significant for the insurance purchasing decision in Smith and Baquet's (1996) study. Their results confirm findings of earlier studies by Just and Calvin (1990) and Edelman et al. (1990) who found that participation in Multiple Peril Crop Insurance is positively correlated with education level. These results cannot be confirmed by the result in our study. Here, personal characteristics of the study population such as age and education do not play a significant role for the participation decision. On the one hand, this might be explained by insufficient experience with crop insurance. On the other hand, educated farmers may see a range of other instruments to increase and stabilize income in their newly-founded enterprises.

Furthermore, Sherrick *et al.* (2004) hypothesize that insurance users are expected to be more experienced and better educated, indicating greater acceptance of insurance use for modern, more sophisticated approaches to risk management. This assumption was not verified by model results. The use of on-farm risk management instruments by operators (RM) was considered to be a relevant factor for the decision for or against the purchase of crop insurance. Farmers who apply risk management instruments are assumed to have fewer reasons to purchase crop insurance as additional risk management. This relationship is confirmed by the negative tendency of the coefficient, although it is not significant. An interpretation in favor of non-significance is contrary to the accepted thought: Farmers who apply risk-management measures know about their efficiency and effectiveness in reducing risks and assess them as non-sufficient. Indeed, they would like to purchase crop insurance in order to back up the remaining residual risk. Another possible interpretation could be that farmers already applying on-farm risk management instruments in the absence of insurance regard insurance as a more efficient instrument of income stabilization compared with on-farm risk management tools.

The literature contains no uniform assessment on the importance of production risk in analyzing farmers' economic behavior. For instance, Hueth and Hennessy (2002) argue that the presence of risk does not explain the tendency to contract, although it is an important consideration in designing optimal contracts once decisions to contract are made. Fraser (1992) assumes that producers who were more risk-averse with respect to their losses would be more likely to participate in insurance programs. The used risk aversion variable (*RATT*) was of significance and shows the expected tendency.

Table 1: Logit model results

Variable	Description	Logit Model Estimate	T-Ratio
CONSTANT		-3.47*	-2.19
AGE	Age of the respondent (0 = younger than 50; 1 = 50 or older)	0.47	0.69
EDUgen	General educational background (0 = no studies and primary education; 1 = higher education)	0.24	0.30
EDUag	Level of agricultural education (0 = no agricultural education; 1 = agricultural education)	0.54	0.74
EXP	Experience with crop insurance (CI) (0,1)	0.50	0.69
RM	Use of on-farm risk management (0,1)	-0.55	-0.66
SPEC	Specialization: 0 = no grain, 1 = grain	1.85*	2.36
AREA	Agricultural area (in ha)	-0.00001	-0.60
RATT	Risk attitude (with respect to income losses)	0.65*	2.09
	Log-likelihood:		-33.00
	Chi-squared:		18.82
	Significance level:		0.016

Note: * statistically significant at the 5% level.
Source: Own estimations.

5. EVALUATION OF POTENTIAL DEMAND FOR CROP INSURANCE

Traditional crop insurance schemes that cover farm yield risk face problems of asymmetric information. Innovative index insurance contracts, such as area yield insurance or weather index insurance, may overcome these problems, but at the cost of generating basis risk for the farmers. The risk-reducing potential of index insurance contracts depends strongly on the extent to which individual farmers are affected by systemic and idiosyncratic risk, respectively. Hence, introducing index insurance schemes into the market must be preceded by an analysis of whether farmers' risk reduction through selected insurance products generates sufficient demand for these products. By applying a utility-efficient programming model (Lien and Hardaker 2001) to two study farms, a range of different insurance instruments will be evaluated.

Expected utility provides a convenient way to represent risk preferences: its basic idea is that decision-makers maximize expected utility. When income increases, its utility increases less than proportionately for risk-averse decision-makers. Hence, utility is an increasing but downward bending function of income. Expected utility estimates can be translated into certainty equivalents (CE), where CE is the inverse of the utility function and represents the monetary value that a person would accept to avoid a specified risk. Knowing certainty equivalent outcomes not only permits the ranking of risky alternatives, but also facilitates estimating risk premiums. CE simultaneously accounts for the probabilities of risky prospects and the preferences for the consequences (Anderson et al. 1977). Each production activity and application of risk management instrument may influence a decision-maker's expected utility. Examining CE is one approach to investigating the magnitude of this influence. The utility-efficient programming model (UEP) integrates the assumptions and constraints of expected utility theory in an objective function that can be described as follows:

$$\max CE = [(1-r)E(U)]^{1/(1-r)}, \qquad (1)$$

where

CE = certainty equivalent, *r* = absolute risk aversion coefficient, and

$$U = 1 - \exp(1-r)z, \qquad (2)$$

subject to

$$Ax \le b,\ Cx\text{-}Iz = uf,\ \text{and}\ x \ge = 0, \qquad (3)$$

where A is a matrix of technical coefficients for all considered activities, b is a vector of capacities, x is a row vector of adjustable variables, C is a matrix of activity net revenues by state of nature, I is an $n \times n$ identity matrix, z is the annual net income in each state, u is a vector of ones, and f represents fixed or overhead costs. The utility function $U(z)$ is positive ($U'(z) \succ 0$), but decreasing ($U''(z) \prec 0$). This corresponds to decreasing absolute $r_a(z) = -U''(z)/U'(z) = r/z$ and constant relative risk aversion $r_r(z) = zr_a(z) = r$. The variable r was set to a value of 2, which represents slight risk aversion.

The activity net revenues for all states, C, represent the uncertainty in activity returns. Therefore, there is no need to assume any standard form of distribution. In our case, suitable values are detrended observations of the time period 1980–2002, treated as states with assessed probabilities.

5.1 Characterization of study farms

A large crop farm (34,000 hectares of sown area) in the grain belt of northern Kazakhstan serves as one of the study farms. From 1999 to 2003, about 74% of the farm income was generated by crop production, 21% by livestock production and 5% by processing. This farm is an interesting object of investigation due to its relatively high yield level, the strong weight of crop production in overall economic performance and the 'organizational history' of the enterprise. After a recent restructuring of the enterprise, the new organizational form is a limited

Table 2: Characterization of study farms

	North-central Kazakhstan	Eastern Kazakhstan
Legal form	Ltd.	Ltd.
Year of foundation	1998	1996
Size (crop area in ha)	34,272	3,100
Irrigated area	-	-
Number of employees (mean 2000-2003)	100	63
Own capital (mean 2000-2003) in '000 KZT	349,397	28,795
Income from (%)		
crop production	85	100
livestock production	12	
Processing	3	
Specialization	wheat, barley	wheat, sunflowers
Average yield power	35	62
Average wheat yield (1999-2003)*	13.9	19.6
Coefficient of variation (1999-2003)	0.117	0.203
Future investment intentions	processing, air operations	–

Source: Farm survey data (Heidelbach 2007).

liability company, where the manager holds more than half of the shares. Several years ago, the farm entity was incorporated in a supra-regionally active holding company that comprised approximately 400,000 hectares of sown area. In the past, production, credit, and insurance decisions were taken by the central planning unit of the holding company. Nowadays the general manager makes most of the important decisions.

The study farm in eastern Kazakhstan is a Limited Liability Company and was founded in 1996. It has 63 full-time employees and comprises 3,100 ha that are to a large extent sown with wheat and sunflowers. The good soil conditions can be attributed to the location of the enterprise at the foothills of the Altai Mountains. Further, the farm has relatively good access to infrastructure important for marketing its products, such as road access and short distances to the processing industry.

Table 2 provides a summary of farm characteristics such as legal form, size, income sources, and crop yield characteristics.

5.2 Description of insurance instruments

Based on Fisher's separation theorem (Fisher 1933), which implies that it is better to diversify through capital markets than through a combination of enterprises, the model includes various insurance and credit activities. In contrast to countries where market-based crop insurance programs are already established and abundant data is available for analysis (compare the studies of Babcock et al. 2005; Bourgeon and Chambers 2003; Miranda 1991; Schnitkey *et al.* 2003), this application requires the pre-formulation and testing of insurance and hedging products before they can be introduced to the risk programming model. The formulation and testing of financial risk management products was carried out by Bokusheva *et al.* (2006). A novelty of this paper is the integration of weather index insurance *based on farm level data*. Skees *et al.* (2001) report a reduction of 29% of the aggregated *regional* revenue risk measured by the coefficient of variation of a portfolio of several crops measured by their regional yield in 17 Moroccan provinces. Vedenov and Barnett (2004) as well as Karuaihe *et al.* (2006) analyze weather-based insurance contracts for *regional* corn, cotton and soybean yields in two U.S. counties and for corn only in South Africa, respectively. The use of regional yields, however, underestimates farm yield deviations and severely biases analysis of yield risk reduction.

In our study we compare the effectiveness of three different insurance types – weather-based index insurance, area yield insurance and farm yield insurance with further financial risk management instruments such as credit, as well as technological risk-reducing tools available to farms in Kazakhstan.

The investigated weather index product is based on precipitation. Annual values of the index xi are computed by the following equation:

$$X_i^{Rain} = W_{May} R_i^{May} + W_{June} R_i^{June} + W_{July} R_i^{July} + W_{August} R_i^{August} + W_{Sept-April} R_i^{Sept-April} \quad (4)$$

where R is the cumulative precipitation, i is a yearlong index and each w represents a weighing factor obtained from linear regressions of the right-hand side variables using farm yields as the dependent variable.

Area yield insurance was formally described by Miranda (1991). Selected products were calibrated for the location of the considered enterprises and included in the model. Area yield insurance products are based on different *underlyings*, i.e., national, oblast and rayon (district) yields. Premium costs and indemnities were estimated based on historical yield and weather data for different strike levels.

Additionally, for rainfall insurance and area yield insurance, conventional farm yield insurance was evaluated. To prevent moral hazard and to obtain a comparable insurance product, the strike level for farm yield insurance was limited to 75% of the expected yield. The following five insurance products were investigated in this contribution: National yield insurance, oblast yield insurance (OYI), rayon yield insurance (RYI), farm yield insurance (FYI), and rainfall-based index insurance (RFI). The estimations are based on yields that were detrended and tested for structural break to establish an input data base that is not affected by technology changes. Strike yields vary between 100 and 75% of the expected yield.

Estimations are restricted to areas with main crops. Special products like potatoes, fruits and vegetables are not considered in the programming model for three reasons: First, their share in total sown area is relatively small. Second, they are only partially marketed and serve, to a large extent, as the basic food supply of farm laborers. Third, since many of these crops are not grown consecutively (i.e., each year) it is not possible to derive statistically firm distribution functions for them.

A range of insurance products with adequate strike levels was selected for stepwise testing in the programming model. Besides risk-sharing instruments such as insurance, on-farm risk management instruments such as irrigation and pesticide application exist. The model takes this aspect into account by determining the utility-maximizing technology.

5.3 Programming model results

Results of the programming model provide answers to the question of what impact do different insurance products have on the utility of a slightly risk-averse decision-maker. A separate regional analysis was a reasonable procedure, because no general recommendations can be derived from the efficiency results.

Results for the study farm in north-central Kazakhstan stress the utility-enhancing effect of insurance, even if optimal on-farm risk management measures are applied (Table 3). If the farmer takes rainfall insurance, for example, the certainty equivalent income can be increased by 5.2% as opposed to a reference scenario without insurance. At the same time, the risk premium[68] is reduced by 5.9% compared to the reference scenario. Alternative insurance products also achieve better results than the reference scenario. Rainfall insurance is followed by farm yield insurance, rayon yield, oblast yield and

Table 3: Certainty equivalents for different insurance product choices – study farm in north-central Kazakhstan (optimal technology)

Scenario	Insurance product	Certainty equivalent (in '000 KZT)	Utility-ranking*	Risk premium
R	- (reference scenario without insurance)	416,070	.948	.128
1	RFI_100	439,003	1	.069
2	FYI_75	434,394	.990	.081
3	RYI_100	434,272	.989	.083
4	OYI_100	427,945	.975	.097
5	NYI_100	425,057	.968	.105

Note: * achieved certainty equivalent as share of maximum achievable certainty equivalent.
Source: own estimations (Heidelbach 2007).

---

68 The risk premium (RP) measures the largest amount of money a decision-maker is willing to pay to replace a random revenue by its expected value, and is assessed as a ratio of the difference between expected income and certainty equivalent to expected income *(risk premium=(expected income - certainty equivalent)/(expected income))*.

national yield insurance. This ranking is in accordance with intuitive evaluations, since the correlation between individual farm and area yields decreases with increasing aggregation levels.

Figure 2 explains how higher certainty equivalents are achieved through insurance products. Indemnity payments compensate losses and increase the net income in unfavorable states of nature. As a result, smoother income distribution increases the certainty equivalent of the risk-averse decision-maker.

Figure 2: Income stabilization through rainfall insurance

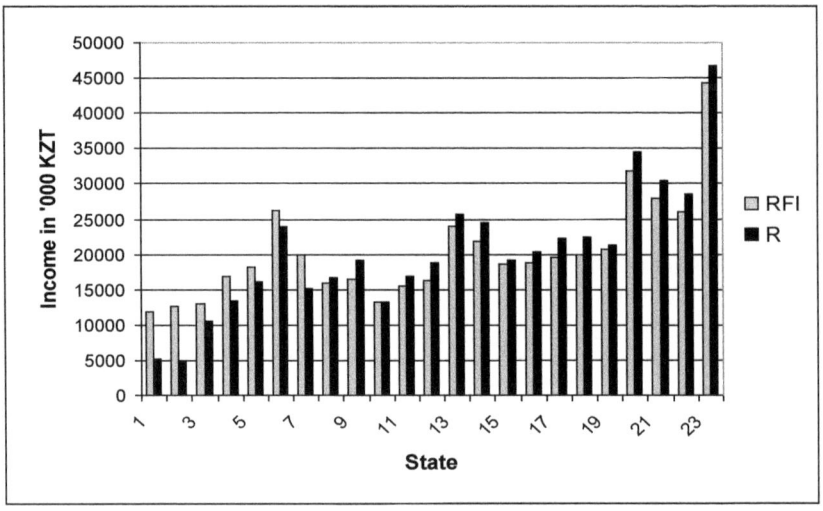

Source: own estimations (Heidelbach 2007).

Extreme losses of up to 24.8% can be caused by the application of a sub-optimal technology compared to a situation where an optimal technology is applied (Table 4). Although the absolute losses are high, insurance can buffer sharp income shocks with an even lower risk premium compared to the results under an optimal technology.

Table 4: Certainty equivalents for different insurance product choices – study farm in north-central Kazakhstan (sub-optimal technology)

Scenario	Insurance product	Certainty Equivalent (in '000 KZT)	Utility-ranking*	Risk premium**
1	RFI_100	312,893	1	.036
2	FYI_75	309,062	.988	.048
3	RYI_100	308,977	.987	.049
4	OYI_100	303,529	.970	.068
5	NYI_100	303,029	.968	.069

Note: * achieved certainty equivalent as share of maximum achievable certainty equivalent.
Source: own estimations (Heidelbach 2007).

The situation is different in eastern Kazakhstan, where income shocks caused by adverse weather conditions are not as extreme as in the northern part of the country. The precipitation is higher and in addition to wheat, sunflowers are grown on a large area. This crop portfolio implies a hedge effect, since wheat and sunflower yields are not strongly correlated.

Table 5: Certainty equivalents for different insurance product choices – study farm in eastern Kazakhstan

Scenario	Insurance products	Certainty Equivalent ('000 KZT)	Utility-ranking*	Risk Premium
R	- (reference scenario without insurance)	27,491	.956	.082
1	FYI_75	28,757	1	.039
2	RYI_100	27,992	.973	.065
3	OYI_100 (wheat), NYI_100 (sunflowers)	27,756	.965	.073
4	RFI (wheat), FYI_75 (sunflowers)	27,111	.943	.094
5	NYI_100	26,668	.927	.109

Note: *achieved certainty equivalent as share of maximum achievable certainty equivalent.
Source: own estimations (Heidelbach 2007).

This effect explains a comparatively lower utility-efficiency of index-based insurance products compared to the northern study farm. Rainfall insurance for wheat in combination with farm-yield insurance[69] for sunflowers results in an even lower CE compared to the reference scenario without insurance (Table 5). For this study farm, farm yield insurance achieves the highest CE, followed by rayon-yield insurance and a combination of oblast and national yield insurance.[70]

6. CONCLUSIONS AND RECOMMENDATIONS

The *supply-side analysis* provided insight into the institutional framework of the crop insurance system in Kazakhstan. It became clear that the institutional preconditions for establishing a proper system are rather weak and do not support the political objective of providing large and small agricultural producers with equal access to crop insurance. Furthermore, the available insurance products are not preventing the problem of asymmetric information to a satisfying degree. Pilot-testing alternative insurance products could supplement specific state legal actions, which should aim at creating incentives for insurance companies to provide high-quality insurance services to all customers.

The *demand-side analysis* is based on the examination of risk reduction of selected insurance instruments and the interpretation of a binominal logit model, which was employed to identify the main factors determining demand for crop insurance. Model results indicate that potential buyers of crop insurance are characterized by a specialization on grain production and a high degree of concern about income losses as a consequence of risk. A product-specific demand analysis could verify the abovementioned results.

The *discussion of potential insurance products* suggests that innovative products such as area yield and weather index insurance are most appropriate for transition conditions. This conclusion is partly underlined by the results of the employed mathematical programming model. However, conclusions regarding the efficiency of considered insurance instruments should be drawn considering regional specifics.

Further research could test the effects by considering alternative utility functions and risk aversion coefficients on income and utility. Furthermore, study farms in

---

69 Correlation tests between weather parameters and sunflower yields did not yield sufficient values for a decision to design weather-based products for sunflowers as well.
70 Oblast yield insurance could not be designed for sunflowers due to missing data.

other regions should be analyzed to derive conclusions on a broader scale. Finally, further production activities could be integrated into the programming model to better evaluate the portfolio effect.

REFERENCES

Anderson, J.R., Dillon, J.L. and Hardaker, J.B. (1977), Agricultural Decision Analysis, Iowa: The Iowa State University Press.
Babcock, B.A. and Hart, C.E. (2005), 'Influence of the Premium Subsidy on Farmers' Crop Insurance Coverage Decisions, Working paper 05-WP 393, Center for Agricultural and Rural Development, Iowa State University.
Bokusheva, R., Breustedt, G. and Heidelbach, O. (2006), 'Measurement and Comparison of Risk Reduction by Means of Farm Yield, Area Yield, and Weather Index Crop Insurance Schemes – The Case of Kazakhstani Wheat Farms', poster paper prepared for presentation at the International Association of Agricultural Economists, Gold Coast, Australia, 12–18 August 2006.
Bourgeon, J.-M. and Chambers, R.G. (2003), 'Optimal Area-Yield Crop Insurance Reconsidered', *American Journal of Agricultural Economics*, 85, 3, pp. 590–604.
Coble, K.H., Knight, T.O., Pope, R.D. and Williams, J.R. (1996), 'Modelling Farm-level Crop Insurance Demand with Panel Data', *American Journal of Agricultural Economics*, 78, 2, pp. 439–447.
Fraser, R.W. (1992), 'An Analysis for Willingness-to-Pay for Crop Insurance', *Australian Journal of Agricultural Economics*, 36, 1, pp. 83–95.
Edelman, M.A, Schmiessing, B.H. and Khajaseh, K. (1990), 'Effects of Disaster Assistance on Multiple Peril Crop Insurance Purchases by Iowa Crop Farmers', Paper presented at the AAEA annual meetings.
FAOSTAT, 'Statistical Database of the Food and Agricultural Organisation of the United Nations', http://faostat.fao.org/, retrieved 24.05.2005.
Fisher, I. (1933), The Theory of Interest, New York, NY: The Macmillan Co.
Heidelbach, O. (2007), Efficiency of selected risk management instruments – An empirical analysis of risk reduction in Kazakhstani crop production. Studies on the Agricultural and Food Sector in Central and Eastern Europe, Vol. 40, Halle (Saale).
Hueth, B. and Hennessy, D.A. (2002), 'Contracts and Risk in Agriculture: Conceptual and Empirical Foundations', in: Just, R.E. and Pope, R.D. (eds.), *A Comprehensive Assessment of the Role of Risk in U.S. Agriculture*, Norwell, MA: Kluwer Academic Publishers.
Just, R.E. and Calvin, L. (1990), 'An Empirical Analysis of U.S. Participation in Crop Insurance', Unpublished Report of the Federal Crop Insurance Corporation.
Khaldi, N. (1975), 'Education and Allocative Efficiency in U.S. Agriculture', *American Journal of Agricultural Economics*, 57, 4, pp. 650–657.
Karuaihe, R.N., Wang, H.H. and Young, D.L. (2006), 'Weather-Based Crop Insurance Contracts for African Countries', Contributed paper prepared for presentation at the International Association of Agricultural Economists, Gold Coast, Australia, 12–18 August 2006.

Lien, G. and Hardaker, J.B. (2001), 'Whole-farm Planning under Uncertainty: Impacts of Subsidy Scheme and Utility Function on Portfolio Choice in Norwegian Agriculture', *European Review of Agricultural Economics*, 28, 1, pp. 17–36.

Miranda, M. J. (1991), 'Area-Yield Crop Insurance Reconsidered', *American Journal of Agricultural Economics*, 73, 2, pp. 233–242.

Mishra, A.K. and Goodwin, B.K. (2006), 'Revenue Insurance Purchase Decisions of Farmers', *Applied Economics*, 38, 2, pp. 149–159.

National Bank of Kazakhstan (2006), 'Cvedenia po ctrachovym organisatiyam Respubliki Kazachstan po costoyainyu na 1 yanvarya 2006 goda (Information about insurance organizations of the Republic of Kazakhstan as of 1st of January 2006)', online at: www. nationalbank.kz.

Nelson, R.D. (1985), 'Forward and Futures Contracts as Preharvest Commodity Marketing Instruments', *American Journal of Agricultural Economics*, 67, 1, pp. 15–23.

Prokhorov, I. (2005), 'Law on Mandatory Crop Insurance', online at: www.zakon.kz, accessed at 11 May 2005.

Schnitkey, G.D., Sherrick, B.J. and Irwin, S.H. (2003), 'Evaluation of Risk Reductions associated with Multi-Peril Crop Insurance Products', *Agricultural Finance Review*, 63, 1.

Sherrick, B.J., Barry, P.J., Ellinger, P.N. and Schnitkey, G.D. (2004), 'Factors Influencing Farmers' Crop Insurance Decisions', *American Journal of Agricultural Economics*, 86, 1, pp. 103–114.

Skees, J., Gober, S., Varangis, P., Lester, R. and Kalavakonda, V. (2001), 'Developing Rainfall-Based Index Insurance in Morocco', World Bank Policy Research Working Paper 2577, Washington, DC.

Smith, H.V. and Baquet, A.E. (1996), 'The Demand for Multiple Peril Crop Insurance: Evidence from Montana Wheat Farms', *American Journal of Agricultural Economics*, 78, 1, pp. 189–201.

Smith, V.H. and Goodwin, B.K. (1996), 'Crop Insurance, Moral Hazard, and Agricultural Chemical Use', *American Journal of Agricultural Economics*, 78, 2, pp. 428–438.

Statistical Yearbook USSR (1985), Moscow.

Statistical Handbook States of the Former USSR (1995), Moscow.

Statistical Yearbook CIS countries (1994), Moscow.

Statistical Yearbooks Kazakhstan (1984, 1987, 1990, 2004), National Statistical Agency, Almaty.

Tazhmakin, D. (2003), 'Oral Information', in: *Astana*, at: 11 September 2003.

Uaisovich, U.A. (2003), 'Oral information', provided by a former inspector of the state insurance system, in: *Esil*, at: 23 October 2003.

Van Asseldonk, M.A.P.M., Meuwissen, M.P.M. and Huirne, R.B.M. (2002), 'Belief in Disaster Relief and the Demand for a Public-Private Insurance Program', *Review of Agricultural Economics*, 24, 1, pp. 196–207.

Vedenov, D.V. and Barnett, B.J. (2004), 'Efficiency of weather derivatives as primary crop insurance instruments', Journal of Agricultural and Resource Economics, 29, 3, pp. 387–403.

Welch, F. (1970), 'Education in Production', *Journal of Political Economy*, 78, 1, pp. 35–59.

Wozniak, G.D. (1984), 'The Adoption of Interrelated Innovations: A Human Capital Approach', *Review of Economics and Statistics*, 66, 1, pp. 70–79.

APPENDIX

Appendix: Trends in crop insurance market development

Oblast	No. of contracts 2005	No. of contracts 2006	Change 2006/2005	Total insured area (ha) 2005	Total insured area (ha) 2006	Change 2006/2005
Akmolinskaya	961	1,837	1.91	1,870,060	3,520,477	1.88
Aktubinskaya	632	638	1.01	243,963	245,789	1.01
Almatinskaya	1,210	1,997	1.65	45,941	21,149	0.46
East Kazakhstan	2,016	5,191	2.57	572,056	370,794	0.65
Djambulskaya	541	724	1.34	46,909	56,975	1.21
West Kazakhstan	521	700	1.34	223,930	330,435	1.48
Karagandinskaya	486	952	1.96	203,549	372,604	1.83
Kysylordinskaya	81	102	1.26	20,721	26,979	1.30
Kostanaiskaya	3,339	3,404	1.02	2,808,782	3,097,948	1.10
Pavlodarskaya	777	791	1.02	473,735	489,024	1.03
North Kazakhstan	2,319	2,303	0.99	1,714,245	1,917,382	1.12
South Kazakhstan	189	369	1.95	2,107	4,662	2.21
Total	**13,072**	**19,008**	**1.45**	**8,225,998**	**10,454,218**	**1.27**

Source: Own formation based on information provided by JSC "Grain Insurance Company", JSC "Trans Oil", and JSC „Victoria", 14 April 2006.

Emerging Markets Studies

Edited by Joachim Ahrens, Alexander Ebner, Herman W. Hoen, Bernhard Seliger and Ralph Michael Wrobel

The Peter Lang series *Emerging Markets Studies* includes works which address opportunities, problems, and challenges of socio-economic development and reform in so-called emerging markets. These comprise middle-income developing and transition economies which are relevant for the world economy due to a large market potential, a favorable or improving investment climate, or due to the availability of important natural resources. Emerging markets have realized or show the potential to generate sustained socio-economic development and growth processes over time.

The volumes in this series seek to address three key questions: What are the determinants of successful socio-economic development, What are appropriate reform strategies to overcome impediments to catching-up processes, and How do politico-institutional factors affect the performance of an emerging economy?

The scope of the series is comparative, institutionalist, and international. The overall focus of all titles is to enhance the understanding of socio-economic catching-up processes and their institutional foundations from a political-economy perspective. Due to the complexity of development processes and policy reform, various methodological tools and academic approaches may prove to be appropriate. Hence the series includes contributions from various disciplines such as economics, political science, or sociology.

Vol. 1   Dominik F. Schlossstein: Institutional Change in Upstream Innovation Governance. The Case of Korea. 2010.

Vol. 2   Manuel Stark: The Emergence of Developmental States from a New Institutionalist Perspective. A Comparative Analysis of East Asia and Central Asia. 2012.

Vol. 3   Bernhard Seliger: The Shrimp that Became a Tiger. Transformation Theory and Korea's Rise After the Asian Crisis. 2013.

Vol. 4   Dirk Johann: The Reconfiguration of a Latecomer Innovation System. Governing Pharmaceutical Biotechnology Innovation in South Korea. 2013.

Vol. 5   Roman Vakulchuk: Kazakhstan's Emerging Economy. Between State and Market. 2014.

Vol. 6   Joachim Ahrens / Herman W. Hoen (eds.): Economic Development in Central Asia. Institutional Underpinnings of Factor Markets. 2014.

www.peterlang.com

www.ingramcontent.com/pod-product-compliance
Ingram Content Group UK Ltd.
Pitfield, Milton Keynes, MK11 3LW, UK
UKHW021822140426
5217IPUK00004B/48